MW00465726

Lean Safety
Gemba Walks

Lean Safety Gemba Walks

A Methodology for Workforce Engagement and Culture Change

Robert B. Hafey

CRC Press
Taylor & Francis Group
Boca Raton London New York

CRC Press is an imprint of the
Taylor & Francis Group, an informa business

A PRODUCTIVITY PRESS BOOK

CRC Press
Taylor & Francis Group
6000 Broken Sound Parkway NW, Suite 300
Boca Raton, FL 33487-2742

© 2015 by Taylor & Francis Group, LLC
CRC Press is an imprint of Taylor & Francis Group, an Informa business

No claim to original U.S. Government works

Printed on acid-free paper
Version Date: 20140711

International Standard Book Number-13: 978-1-4822-5898-1 (Paperback)

Library of Congress Cataloging-in-Publication Data

Hafey, Robert B.
  Lean safety gemba walks : a methodology for workforce engagement and culture change / Robert Hafey.
    pages cm
  Includes bibliographical references and index.
  ISBN 978-1-4822-5898-1
  1. Manufacturing processes--Safety measures. 2. Industrial safety. 3. Industrial management. 4. Organizational change. I. Title.

TS183.H337 2015
670--dc23   2014024629

Visit the Taylor & Francis Web site at
http://www.taylorandfrancis.com

and the CRC Press Web site at
http://www.crcpress.com

Dedication

To those who by working with their hands provided me the material
and inspiration to write this book. May the material in this book
help keep them and all who work with their hands safe.

Contents

Acknowledgments

My personal Lean Safety journey began when I was hired to work at U.S. Steel Corporation at the impressionable age of eighteen. Walking into a large steel mill where red hot metal was being processed on very large equipment was scary. The plant I worked in processed 4-inch square steel billets into a glowing red hot steel rod that snaked its way through multiple roll stands, which reduced and sized the hot metal into the finished product. My career has mirrored the snaking path of that hot rod for it has taken many turns as I have been formed into someone who has a passion for both continuous improvement and safety. It has been a 40-year Gemba walk through life. The distillation of those work-life experiences and the influence of all of the people I worked with have had a profound influence on how I view work cultures.

I learned as much from the negative influences as I did from the positive. As I observed managers who were driven by a personal desire to grow sales and profits, I watched work cultures suffer. Conversely, as I observed leaders invest in and grow their people, I witnessed business cultures flourish. People are the game changer. They are the one common ingredient in every business. Work cultures determine the long-term success of a business and they are a mirror reflection of the people in charge of the business. By peering into the company culture mirror, a manager can quickly assess if he or she is a leader or just a manager. Managers create work cultures that have parent–child relationships. Leaders enjoy work cultures that have adult relationships. Leading is fun and renewing. Managing is just a tiring job.

Thank you to all the leaders who continue to provide me with challenging and fun work-life experiences by inviting me into their facilities to share my passion for Lean Safety. It reinvigorates me every time you afford me the opportunity to engage your staff in adult conversations about making work safer and easier. Thank you to the organizers of the many Lean Safety events held around the world and to all of the workshop participants. As my life's journey continues to snake along, each of you has influenced me as a person. Life is a shared Gemba walk with family, friends, and acquaintances. I am glad all of you have been part of my life.

Introduction

What is your passion? No matter what your age, having something you are passionate about allows you to add value to the world and feel good about yourself. As a member of the baby-boomer generation, I am witnessing many friends and acquaintances retire. Recently, I emailed a business associate and his Microsoft Outlook "out of the office" response noted, "I will be out of the office FOREVER." The question he and others have to answer is what are you going to do with forever? Just as life has cycles, so can our passions—they can come and go or last a lifetime. For many, their work is a lifelong passion. When they stop working, they seem temporarily lost. Relationships with co-workers along with the sense of value they brought to the world end abruptly. They struggle to find a new way to make a difference in the world. At times, they blame others for their state of confusion and relationships suffer. However, passion is personal. It is a self-discovery process that is ours alone and it can be a painful process until we make the discovery.

Traditional retirement activities like sports, outdoor activities, and travel (not to be confused with tourism) fill some of my days. I am thankful I live in the Chicago area where, because of the weather, I am unable to play golf for five months of the year. After all, I only play golf to stay humble! Outdoor activities like hiking help keep me fit and are important for that reason alone. Travel, which provides uncertainty and unanticipated interactions with new people, places, and cultures, can be a real joy. But I knew before I left my full-time position those types of activities would not allow me to feel as if I was still engaged and making a difference in the world. Therefore, I never intended to "retire" in the historical sense of the word. It has been four years since I stopped working full time and started my new career as a part-time consultant, and they have been four of the best years of my life. Having passion for the continuous improvement of safety allows me to share it with others and in return build new and lasting relationships. Often when beginning a workshop I ask the attendees to introduce themselves—name, company, position, and one thing they have a passion for outside of work. When respondents do indeed have a passion for something, I love to watch their eyes and body language as they describe whatever it is. I see myself in them when I talk about my new career or one of my other passions.

What does this have to do with Lean and safety? True Lean leaders challenge and grow people (engagement). They provide their reports with the gift of time (empowerment) to improve the business. This has a secondary effect. It allows their reports to grow as individuals. Empowerment leads to engagement and when people are engaged, they find passion. A business full of passionate people is a competitive weapon. Each of us has a responsibility to empower and grow those who surround us in life. To watch spouses, children, co-workers, friends, and reports grow as individuals and develop their own individual passions in life is one of the joys of life. I hope you have many passions and are inspiring others to find theirs.

My passion for Lean Safety began in 2008 and really took off at the start of 2010 when the book I authored, *Lean Safety—Transforming Your Safety Culture with Lean Management,* was published. Since then I have had many opportunities to travel and share my Lean Safety message. In that first book I attempted to help the readers make the connection between my own positive safety experiences and the continuous improvement philosophy titled Lean. I wanted them to understand they could take the "mean out of Lean" and at the same time make work safer and easier. Management teams around the world have gotten it wrong all too often. Their belief that Lean, the continuous improvement philosophy based on the Toyota Production System, is a cost-reduction rather than a customer-focused program had alienated their employees. It is impossible to truly engage the hearts and minds of a workforce in continual improvement if they believe the result of their efforts will be the loss of their employment. This misuse of the Lean philosophy has created a deep and widening trust gap between hourly employees and management.

To narrow or bridge this gap, one has to find common ground on which to begin a new Lean implementation dialogue. That common ground is safety. Safety is a rallying point for all stakeholders. Unions, managers, front line supervisors, and hourly employees will all get in line and support safety. One might think this would occur naturally, but it doesn't because safety in most plants is based solely on compliance to regulatory agencies like OSHA. To ensure compliance, senior managers hire environmental, health, and safety (EHS) professionals who then manage the safety program and its lagging indicators in a top down directive fashion. Using a Lean term, safety professionals "push" safety and are seen as the enforcers of safety by the workforce. This legacy safety system, which also uses discipline for safety infractions, relies on fear and intimidation to help ensure compliance. So, how can someone turn a legacy safety program into something that will build trust and become the foundation of a company's Lean efforts? Simple—engage and focus the employees of a business on the continuous improvement of safety.

Over the last four years, I have had the opportunity to make a difference in the world by changing how the people I have interacted with view workplace safety. I have given presentations, conducted workshops, and facilitated Lean Safety Gemba Walks and safety kaizen blitz events. As a result, the attendees leave understanding that even though safety compliance requirements are always going to be present, they can and should view safety as just another part of a business's continuous improvement program. The integration of Lean and safety is a natural, almost organic process. A common example of this amalgamation is a safety kaizen blitz. The kaizen blitz (a team-based multi-day rapid improvement event) is a Lean tool used to drive cycle time out of a process like a machine changeover. Using this same approach to target the reduction of soft tissue injury risks in a work process, rather than cycle times, sends a different message to the workforce. Removing the stopwatch from the event eliminates the symbolic threat of job loss. Participants in these safety kaizen events walk away with a new approach to help them understand both safety and continuous improvement. They view them as a unified approach to continuous improvement for making work safer and easier as well as reducing cycle times.

The traditional kaizen blitz, used to reduce business process cycle times, has long been criticized because the improvements or gains made during kaizen events often are not sustained. To those critics I say, so what? Mature Lean thinkers understand that for a business to become Lean it must impact how people think, act, and interact. This cultural or people side of Lean is the difficult "row to hoe" on anyone's Lean journey. When businesses hold multiple kaizen blitz events a week, they are doing so to improve business process cycle times and to win the hearts of their employees. They understand the biggest benefit of any kaizen event is not the process you improve—it is the minds you change.

Hourly employees, those who do the real hard work for a business's customers and are the least empowered employees in any business, need to understand how "Lean thinking" benefits each of them individually as well as the business. Engaging your employees in kaizen blitz events that focus on the reduction of injury risks sends a completely different message. Skeptical hourly employees, and most of them are, will quickly become Lean champions when they see management giving focus to their safety while seeking cycle time gains. Remember, people do not care how much you know until they know how much you care. Make work safer, win their hearts, and then move your Lean efforts forward.

However, before suggesting a safety kaizen event, I first take individuals on a Lean Safety Gemba Walk. Gemba, a Japanese word, is part of the Lean community vernacular. It means the factory floor, the hospital floor, the restaurant kitchen, the warehouse aisles, a construction site, or anyplace "where the work is done." The word Lean has become synonymous with continuous improvement. Therefore, a Lean Safety Gemba Walk is a walk through the work area that focuses on the continuous improvement of safety. The walks I have led have ranged from one-on-one events with senior leaders to guiding large groups of workshop attendees on a journey that changes how they view safety. Lean Safety Gemba Walks have nothing to do with compliance safety. Rather than focus on "things," the sole focus is on the individuals doing the work. By watching the actions required to complete work tasks, it is easy to identify improvement opportunities that will make the work safer and easier. When conducted in a respectful manner, by a skilled facilitator, these Safety Gemba Walks have a dramatic long-lasting impact on the safety culture of a business. They engage managers and hourly staff in the continuous improvement of safety. Employees now have a chance to make a difference in their safety culture rather than just be compliant with the rules. Everyone involved in business change wants the "what's in it for me" question answered. It is not the fear of change, but the unknown result of the change, that keeps people from fully participating in Lean efforts. When shop floor employees and their managers engage in genuine workplace safety improvement, trust building begins for a safer workplace in a "what's in it for me" outcome beneficial to all. Lean Safety Gemba Walks are a business culture game changer.

This book contains a collection of Lean Safety Gemba Walk case studies based on my experiences over the last four years. As the stories unfold, the reader will be transported to the Gemba and begin to see safety differently just as those who physically participated. By sharing these success stories, I can have an even greater impact on the safety of those to whom my first book, *Lean Safety*, was dedicated—"those who work with their hands and attempt to stay out of harm's way." Writing this book is the next step on my journey to change the world—or at least how the world views workplace safety.

Chapter 1

The Common Objective—Impact the Culture by Building Trust

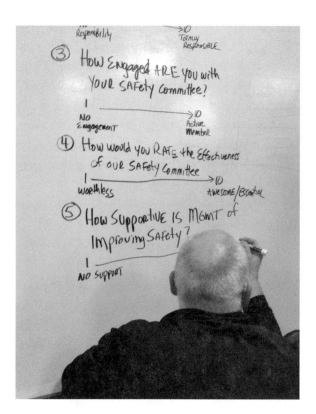

Lean and Lean Safety follow the same path to success. The definition I most often use to describe a work culture is "how people think, act, and interact." Clearly, then, the path to success has to do with changing the work culture.

Or, to say it another way, the objective is to get everyone to think differently. It is common knowledge in the Lean community and often quoted by Lean proponents and critics that successful Lean transformations are few and far between. Why is that? Most often because Lean is thought of as a program, rather than a management philosophy, and used by management to reduce labor costs. The second and equally harmful reason is that management fails to have the will to stick with Lean. It takes a strong will and a strong leader to start and stick with a plan to redirect a business culture for it is always a multi-year effort that requires honest discussions and, often, difficult decisions. Lean doesn't fail—it is always management that fails. In these failed attempts, Lean is viewed as yet another "flavor of the month" program that yields no long-term benefits to the business. In both cases, where the Lean objectives were wrong or a lack of will existed, the trust building required to move a culture forward did not exist. Without trust, culture change is a bust.

Management has always owned the culture that exists within their business. Their actions, or lack of action, either build or tear down the level of trust. There are no neutral actions. Lean thinkers often say, "the Gemba is a reflection of management." A walk through a plant is just like an open window through which you can view a company culture. The level of trust that exists can be gauged quickly and simply by the level of engagement witnessed when a visitor approaches and attempts to talk with the workforce. There are many parallels between Lean and safety culture change within a business. Trust building is also the key to safety culture change, yet one aspect of current compliance-based safety cultures that gets in the way of trust building is the use of discipline for safety infractions or noncompliance. Eliminating the use of discipline is an opportunity for management to make a stand that will redirect the safety culture, but it is rarely done for reasons I will discuss later in the book. Another problem for the compliance safety community is they often do not know how, or feel they have approval, to try to redirect the safety culture because they are so immersed in compliance safety.

A large percentage of those who attended the Lean Safety events I have facilitated over the last four years have been safety professionals, whose sole focus is safety, or individuals with some safety responsibilities in addition to other responsibilities, that range from environmental, health, operations, and HR. They often do not have a clear understanding of Lean because they have not been included in the Lean activities that take place where they work. Adding to their confusion is the fact that Lean practitioners often use Japanese words to convey their message about Lean tools and concepts. I have found that I have to do some trust building with these safety professionals early on in a workshop for later that same day I am going to challenge some of their beliefs. Early on the first morning of a workshop, I ask each table grouping to write a definition of Lean that they will share with the other workshop attendees. Then, as I present the workshop material, the Lean terminology, including Japanese words, is clearly defined for the attendees who lack understanding. In this way, I try to assimilate them slowly into the Lean community by removing the fear all of us feel when exposed to new things that we do not fully understand. I have to earn their trust if I hope to touch them deeply enough so that they will return to work and approach safety differently than they have in the past.

Occasionally, it is not possible to engage a safety professional despite my facilitation skill set. In one case, a company that belonged to a consortium was volunteered by its operations manager to be one of the sites the workshop attendees would visit for a Safety Gemba Walk the day after the classroom portion of the workshop. In a pre-event planning phone discussion, which included the consortium facilitator and the site safety manager, I listened to a defensive safety manager who felt he had been ambushed by the operations manager who had volunteered the site. He was anything but cooperative and when asked if he was going to attend the first day of the workshop he was noncommittal until the very end of the conversation when he said he might be there. He did indeed show up but was guarded and defensive during the whole day despite my attempts to make a connection with him. It didn't get any better when we visited his plant the following day. He quickly made it clear he was in charge of safety by the way he directed the company's safety committee members. He was obviously very insecure and felt that his top-down controlling way of dealing with people was the best approach for safety management. My humanistic engaging approach to safety was in direct conflict with his top-down directive "I am in charge here" style of leadership. I was in his house and I was about to deliver a different message than his—I understood why he felt threatened. My hope was that the results of the Safety Gemba Walk would begin to change his thinking.

Those who joined me on the Safety Gemba Walk were fitted with headsets while I had a headset fitted with a microphone. A Safety Gemba Walk is a trust building activity and I was very interested in observing the culture on the Gemba. Did it mirror the safety manager or was he an anomaly? As we walked, I approached individuals engaged in work. I introduced myself and asked their names. I next explained why we were there and our objective—to make work safer and easier. I asked for their permission to have the team observe them and made it clear that we were very interested in their ideas as well—after all, they were the process experts! As we moved from workstation to workstation, it became clear to everyone that this walk was unlike any past safety walk in which they had participated. We did not look for OSHA noncompliance or internal safety regulation violations. We only watched people work.

Near the end of the walk, I observed a woman who was bent over a metal housing trying to install a drive unit using bolts and nuts. It appeared to be a frustrating job for the holes in the housing and the drive unit did not align properly. After my introduction and explanation of why we were observing her work, I commented that she was permanently bent over with her back and neck out of neutral while working. I asked her if her back ever hurt. She looked directly into my eyes and said, "My back hurts every day." At that moment, she had just revealed to me the key to earning her trust. I asked if she and her co-workers had made any changes to the work process to make the work easier. She referenced a tool someone had made to help align the bolt holes. I then asked if a safety kaizen team were formed and their objective would be to both keep the assemblers in an upright position and make the assembly process easier, would she participate on that team. She quickly replied, yes, if it would make her work easier. While I was talking with her, all of the others via their headsets heard everything I said. I was wondering what the safety manager was thinking. This is not the type of safety he practiced. Because he was a compliance safety manager, on his safety walks he would only look for violations against the standards. Stopping to watch someone work and engaging them in a safety improvement discussion was a foreign concept to him and almost everyone I have taken on a Safety Gemba Walk. What I was doing was engaging people in discussions about making their work safer and easier. I was laying the foundation for trust building that would occur if the kaizen event were scheduled and held. It is such a simple concept, yet managers walk through their facilities day after day with blinders on. They see people working as people just doing their job.

After our walk, we returned to a meeting room where the walk participants, who had been recording all of the opportunities for improvement we observed, read them aloud so they could be recorded on a flip chart. The safety manager offered to operate the flip chart to record the opportunities. As soon as he started to write, I understood why he volunteered. He did not attempt to record the details of what was being said but instead recorded small fragmented notes that did not truly reflect the opportunities noted. It was his last chance to show me he was in charge. He seemed tense and a little upset because to show any excitement about the opportunities would have been an admission that he was in a sense wrong in how he managed safety in the past. Willfully bypassing this opportunity to be a creative leader in order to maintain the status quo was his decision. He was, I am sure, a very good person. He was a product of his compliance-based safety upbringing. I understood he had a compliance job to do and that he must continue to do it. That is what he was paid to do. This was his first introduction to continuous improvement versus compliance safety. I hope that later the Safety Gemba Walk experience caused him to reflect on the benefits of engaging people and building trust. I understood that proactive safety improvements identified and implemented because of a Lean Safety Gemba Walk support and improve a company's compliance safety program. That is what I hoped he would eventually understand—and I always have an over-abundance of hope.

Chapter 2

Compliance-Based Safety— Not Good Enough

That's right—compliance-based safety programs are not enough to prevent injuries. I think everyone understands that—even the people at OSHA. Workplace injury reduction driven by compliance to OSHA has been impressive. Here is some supporting information from the OSHA website.

In 1970, an estimated 14,000 workers were killed on the job—about 38 every day. For 2010, the Bureau of Labor Statistics reports this number fell to about 4,500 or about 12 workers per day. At the same time, U.S. employment has almost doubled to over 130 million workers at more than 7.2 million worksites. The rate of reported serious workplace injuries and illnesses has also dropped markedly, from 11 per 100 workers in 1972 to 3.5 per 100 workers in 2010.*

We all owe a tip of the cap, or safety helmet, to all the people at OSHA, past and present, for the job they have done to protect all of the workers in our country. It is very different in other parts of the world.

A few years ago I gave a presentation and facilitated a workshop at a safety conference in Donetsk, Ukraine. The company that invited me to attend the safety conference it sponsored employs around 130,000 people. This company began, and continues to grow, by purchasing old Soviet-era coal mines and steel mills. As it attempts to westernize these facilities so it is capable of competing with its Western European counterparts, it faces huge safety challenges. Often no formal safety program exists in the newly acquired businesses and the safety professional have to find ways to quickly make a difference. I will expand on my experiences in the Ukraine in an upcoming chapter.

The Ukraine is a relatively new independent country and it is still sorting out and stabilizing its political systems. Therefore, safety regulations enforced by a government agency, like OSHA here in the United States, are nearly nonexistent. If businesses are not policed by a government agency, what drives business owners to comply with safety regulations?

It can be external pressures like a desire to be globally competitive. In the global business arena, a business's safety record and safety culture are put on the world stage. To compete globally, it may have to adhere to standards higher than those in the country in which it operates. In other cases, a supplier or customer might set the expectation for improving safety. One client of mine was at risk of losing a very large customer due to its poor safety record. This drove a corporate effort across all of its sites to reduce its recordable injuries. Another driver of safety change is workplace tragedy. A few years ago, I gave a Lean Safety presentation at a trade association conference. Another presenter was the president of a fabrication business in the Midwest. She talked about losing a brother who was killed in an

---

* http://www.osha.gov/Publications/all_about_OSHA.pdf

industrial accident and the lasting personal impact it had on her. Her view of the importance of, and passion for, workplace safety in her business was changed the day she lost her brother. Nevertheless, whatever drives safety improvement, it is still a drive toward meeting regulations by enforcing compliance—it is compliance-based safety. In addition, remember compliance-based safety is never going to be enough to prevent injuries.

A common characteristic of compliance safety programs is that they are top-down directive programs. This should not come as a surprise to anyone because it all starts with OSHA—a regulatory agency in the United States. Most developed countries have an equivalent agency that defines and enforces the safety regulations for different industry segments. Regulate is defined as "to govern or direct according to rule or to bring under the control of law or constituted authority." Simply stated, OSHA is the enforcer of the safety rules and regulations to which companies must comply. It is the safety police. When any process starts at the top as a directive top-down activity, it cascades in that mode until it reaches the last step in the process. An example is the compliance safety walks that are conducted in most facilities.

Safety walks are an integral part of most compliance safety programs. Just search the Internet for "safety walk" and you will find free downloadable forms that you can use to conduct a traditional safety walk. Safety walks, which are really safety audits, have been used for decades. An individual, or group of individuals, walks through a facility and audits the current state condition against the OSHA standards. They are inspecting the facility to see if it is in compliance with OSHA standards. The auditors focus their attention on "things." If you read through safety walk audit forms, you will find words like *stairs, extinguishers, machines, guards, switches, electrical wires, signs, racks, aisles*, etc. In addition to the internally mandated safety walks, a company's insurance carrier may ask to visit the site and conduct a safety audit. Its walk through will mirror the one previously described. All of these compliance-driven safety walks focus everyone's attention on "things" and fail to look at the people doing the work. If the individuals are observed at all, it is only to ensure compliance. For example, are they wearing their personal protective equipment (PPE)? This type of safety walk may help maintain compliance but it does nothing to move the safety culture of your business forward. It actually keeps your safety program anchored in the past. The strongest anchor rope of compliance safety is the use of discipline for safety infractions.

Compliance-based safety programs, and many safety professionals and managers, rely on discipline to enforce, and reinforce, compliance to the rules. Companies that use discipline routinely have a parent-child work culture. Top-down directive actions drive the activities within the business and people are afraid to take actions on their own. Managers make statements like, "We need to send a strong message to all of our employees," when deciding on the severity of the discipline being considered for a safety rule violation. Their approach is to use fear and intimidation to gain compliance. If a management team is happy with a mediocre business, and a safety program based on compliance alone, they can and should continue to drive fear through the workplace by using discipline. However, if they are trying to build a world-class organization, they must drive fear from the workplace as noted by W. Edwards Deming in his 14 Management points.

William Edwards Deming (October 14, 1900–December 20, 1993) was an American statistician, professor, author, lecturer, and consultant. He promoted the Shewhart Cycle "Plan–Do–Check–Act" named after Dr. Walter A. Shewhart so often that it has also been called the Deming Cycle, but not by him. He is best known for promoting his management method called 14 Points, which is based and derived on a system of thought called the System of Profound Knowledge, consisting of four components: the appreciation of a system, understanding of variation, psychology, and a theory of knowledge. These components work together and should not be separated.*

If you are not familiar with Deming, take your time to read the following. These are golden nuggets of advice for Lean leaders.

> W. Edwards Deming offered 14 key principles for management to follow for significantly improving the effectiveness of a business or organization. Many of the principles are philosophical. Others are more programmatic. All are transformative in nature. The points were first presented in his book *Out of the Crisis*. Below is the condensation of the 14 Points for Management as they appeared in the book.
>
> 1. Create constancy of purpose toward improvement of product and service, with the aim to become competitive and to stay in business, and to provide jobs.
> 2. Adopt the new philosophy. We are in a new economic age. Western management must awaken to the challenge, must learn their responsibilities, and take on leadership for change.

---

* http://en.wikipedia.org/wiki/W._Edwards_Deming

3. Cease dependence on inspection to achieve quality. Eliminate the need for inspection on a mass basis by building quality into the product in the first place.
4. End the practice of awarding business on the basis of price tag. Instead, minimize total cost. Move toward a single supplier for any one item, on a long-term relationship of loyalty and trust.
5. Improve constantly and forever the system of production and service, to improve quality and productivity, and thus constantly decrease costs.
6. Institute training on the job.
7. Institute leadership (see Point 12). The aim of supervision should be to help people and machines and gadgets to do a better job. Supervision of management is in need of overhaul, as well as supervision of production workers.
8. Drive out fear, so that everyone may work effectively for the company.
9. Break down barriers between departments. People in research, design, sales, and production must work as a team, to foresee problems of production and in use that may be encountered with the product or service.
10. Eliminate slogans, exhortations, and targets for the work force asking for zero defects and new levels of productivity. Such exhortations only create adversarial relationships, as the bulk of the causes of low quality and low productivity belong to the system and thus lie beyond the power of the work force.
    – Eliminate work standards (quotas) on the factory floor. Substitute leadership.
    – Eliminate management by objective. Eliminate management by numbers, numerical goals. Substitute leadership.
11. Remove barriers that rob the hourly worker of his right to pride of workmanship. The responsibility of supervisors must be changed from sheer numbers to quality.
12. Remove barriers that rob people in management and in engineering of their right to pride of workmanship. This means, inter alia, abolishment of the annual or merit rating and of management by objective.
13. Institute a vigorous program of education and self-improvement.

14. Put everybody in the company to work to accomplish the transformation. The transformation is everybody's job.[*]

Great leaders understand that building trust is the key to business improvement. They also understand every action they take either builds or tears down the level of trust. Discipline is a trust killer.

In my workshops, I ask if anyone, as an adult, has tripped and fallen to the ground. When I ask those who have raised their hands what was the first thing they did after falling, they laugh and agree it was to look to see if anyone had seen them fall. If indeed they were observed falling, they also agreed the second response was to call out, "I am okay." When queried as to why this was their reaction to falling they make statements like, "I was embarrassed." Individuals who are injured on the job also are embarrassed and suffer physical pain. I believe that is enough. Making then feel worse, or making an example of them, by issuing discipline is of no value to a business today. Rather than issuing discipline, and killing trust, build trust by engaging the injured in defining and implementing work process changes that will prevent the possibility of someone being injured while performing the same task in the future. Make accident investigation meetings continuous improvement meetings. Focus on the "what" and "why," and not the "who." An example I reference in my workshops has to do with an individual whose unsafe act resulted in a minor injury and who then freely admitted he was injured because he failed to follow the lock out tag out procedure that was hanging on the machine he was operating. In many companies, failure to follow lock out tag out procedures will result in discipline—probably time off. But what if rather than issue discipline, the individual was engaged in implementing the corrective actions that resulted from the accident investigation? So that rather than feeling bad, he felt good because he was involved in helping to ensure that no one was ever injured performing this work task again. This new intentional response to injuries will start to create an adult workplace because trust is extended by management with the hope of earning it in return.

Discipline is woven into the fabric of compliance-based safety programs and used as a tool by those who manage them. Insurance carriers even ask their clients to provide discipline records as a way of ensuring a company is serious about safety! The use of discipline is a very contentious subject in the compliance-based safety community and I have had some emotional

---

[*] https://www.deming.org/theman/theories/fourteenpoints

responses when I suggest everyone should stop using discipline. What I understand, and what they may not, is that a safety culture cannot be changed if discipline is used—period. Any culture change is a trust building exercise and management has to give trust to earn it. My goal is to change how the world views safety. I want everyone to understand that safety can be a continuous improvement activity and people cannot be living in fear if you want them to participate with their hearts and minds.

Leaders focus on safety because it is the right thing to do or because they are forced to focus on safety by an external influence like OSHA, global business demands, their customers, or serious injuries or deaths that occur within their company. No matter what drives them, I want them to take safety to the next level. I want to challenge all business leaders to do the right thing—to go beyond the compliance requirements and provide the resources and their own personal energy to make the continuous improvement of safety an integral part of their safety program.

One such leader with whom I work had the chance to meet and work for a short time clearly understood the difference between top-down directive safety and the continuous improvement employee engagement style of safety. He was intent on creating a training program centered on "coaching for safety." I first met him when I visited his facility at the request of the company's global safety manager. He had asked me to visit and meet with the site maintenance manager to offer suggestions to a safety issue related to loading castings into a CNC lathe. When I arrived at the plant, the maintenance manager took me directly to the plant manager's office prior to our assessment of the equipment loading process. The plant manager shared his desire to impact the safety culture by taking it beyond compliance and asked me to think about putting together some material for a workshop that would accomplish that goal. Over the next few months, which were filled with international travel for both of us, we corresponded back and forth on the topic of coaching for safety. Then in early December he attended a public Lean Safety workshop I facilitated to help him better understand my approach. Shortly after the workshop, he invited me to his facility for a final discussion on the topic. I use the word final because during our meeting he told me he was retiring at the end of the year. I responded with, "Twelve months from now?" "No," he said, "in two weeks!" He was excited about the imminent change and yet he worked up to the end to ensure his vision for the new safety culture carried forward by introducing me to the manufacturing manager who would be my contact when we finally scheduled the workshop. A few weeks later when I sent an email to my new company contact, I copied my initial contact, the now retired plant manager, thinking he would still receive his emails. The Microsoft Outlook "out of office" response I received from his mailbox was, "I will be out of the office FOREVER!" Maybe so, but his influence made a real safety difference after his retirement. He understood compliance safety did not go far enough to engage the workforce in proactive safety improvement. Later in the case study section of this book, I will continue the story of the workshop on coaching for safety.

Chapter 3

Behavior-Based Safety versus Lean Safety

I have to confess I am not a safety expert or a safety professional. I am a passionate operational continuous improvement (Lean) professional who wants to change how the world views safety. This has both an upside and a downside. On the positive side, I am not encumbered by being part of the compliance safety community that is so stuck on the compliance side of safety. That is what they are educated to do and what they are paid to manage in their places of employment. They are a required resource in businesses around the world and I am sure most of them do a very effective job. The problem is they have difficulty seeing or have not been exposed to the opportunities presented by continuous improvement safety. It is foreign to many of them because they have not been asked, or required, to lead continuous improvement safety activities in their business. Instead, they were hired, and are paid, to ensure compliance. Another positive is that because I am a Lean thinker, I question everything. To me, a process is a process. Any process can be process mapped, or taken apart, and put back together in a better way. Safety processes are just as relevant a target for this type of improvement activity as are production or paper flow processes in a business. An example is using process mapping to first fully understand and then improve the incident/accident investigation process.

On the downside, some in the safety community see me as an outsider because I have no safety credentials—no CSP (Certified Safety Professional) or any other letters appear after my name. I have submitted call-for-presenter forms for the NSC (National Safety Council) and the ASSE (American Society of Safety Engineers) annual conferences and have never been accepted. I know the people selecting the presenters are probably safety professionals and they must question how someone from outside their safety community could bring value to their event. When I facilitate public Lean Safety workshops, the attendees are always a mix of operational/Lean and EHS professionals. Therefore, I quickly acknowledge right up front the fact that I have two audiences in these public workshops. I also point out to these two groups of attendees they have something in common—the well-being of the people employed in their respective businesses. With that common goal in mind, I guide them through the workshop. During the workshop, I focus them on the requirement to change safety cultures and how to positively impact them. Often a safety professional in the workshop will ask if what I am talking about is behavior-based safety (BBS). My response is no. I am not trying to change people's behaviors; I am trying to make their work safer and easier while at the same time earning their trust. If I can earn their trust and engage them in the safety program, they will change their own behaviors.

Behavior-based safety is a topic that has been around for a long time. BBS originated with the work of Herbert William Heinrich. In the 1930s, Heinrich, who worked for Traveler's Insurance Company, reviewed thousands of accident reports completed by supervisors and from these drew the conclusion that most accidents, illnesses, and injuries in the workplace are directly attributable to "man failures," or the unsafe actions of workers. Of the reports Heinrich reviewed, 73% classified the accidents as "man failures"; Heinrich himself reclassified another 15% into that category, arriving at the still-cited finding that 88% of all accidents, injuries, and illnesses are caused by worker errors.[*]

Here are some excerpts from an article in *EHS Today* titled, "Behavior-Based Safety: Myth or Magic?"

"When it was introduced, behavior-based safety (BBS) was seen as a magic panacea for everything that ailed safety programs. It was the Swiss Army Knife of safety programs. It could take care of everything," says Ron Bowles, director of operations for Portland, Ore.-based Strategic Safety Associates. "Now people realize that it is just one tool and more are needed."

Decades after the initial launch of BBS programs, the process has lost favor with many safety managers, who claim the cost—such programs can be expensive—and the long-term results are not what they expected.

Some experts argue that expectations for BBS were unrealistic from the start, while others believe the process has been corrupted at some companies, transformed into an auditing program that assumes a "blame the employee" attitude about safety failures. "Behavior-based safety makes the assumption you know what behaviors you should be doing," says Robert Pater, managing director of Strategic Safety Associates. "It assumes you know what to do and need to be reminded to do it."

Not surprisingly, that approach failed at many companies, says Larry Hansen, CSP, ARM, author and principal of L2H Speaking of Safety Inc.

"My intro to behavior-based safety was being asked by my employer at the time to go to an Indiana food distribution company to analyze the safety program," remembers Hansen. "At 9 a.m., I walked in the door and the general manager said, 'Stop right there. I just bought a gun and the next SOB who mentions behavioral safety...'"

---

[*] http://en.wikipedia.org/wiki/Behavior-based_safety

Hansen said the company had spent hundreds of thousands of dollars on a behavior-based safety program and it had failed miserably. "It never had a chance," he says. "There was a poor manager and a sick organization. They bought into it because they thought it said what they wanted to hear about the cause of incidents, what I call PDDT: people doing dumb things. In other words, employees are the problem and a BBS program can 'fix' them. It's a core misconception that leads to failure."*

I would guess, or maybe just hope, the BBS consulting firms no longer use terms like "man failures" when selling their programs to management teams, yet that is still the core concept of BBS. As an outsider, it seems like BBS is marketed as a "quick fix" program that can be sold to management teams struggling with high injury rates and ineffective safety programs. Large consulting firms that want to take up residence in their clients' facilities have to create big expansive programs to sell. BBS fits that model. The BBS model allows them to place their consultants for an extended period in a facility in order to maintain their mass. If you explore the BBS topic on the Internet, you will certainly find conflicting opinions, like the article excerpts above, about the value of its usage. BBS detractors feel it is a very costly approach that has a minimal impact in the end, while its supporters, mainly those selling BBS, are quick to point out its financial payback. The core problem I see is that BBS only focuses on one small part of the overall safety process when the management of a business is a myriad of processes that are all intertwined. Addressing one symptom of a broken management system does not fix the whole. Will you see improved short-term safety metric results if you implement BBS? Probably. BBS temporarily redirects the compliance-based safety culture and reduces the injury rate. You get what you measure is an old saying that is applicable in this case. Will you see long-term safety culture change? Not likely, because the focus of BBS is the person and not the process. As a Lean thinker, I am a firm believer that the process, not the person, is the problem. BBS does not dig deep enough to get to the underlying causes of a broken safety culture. It only treats a symptom of that broken safety culture—injuries. I think the underlying reason why BBS is ineffective over the long run is that it is an offshoot of traditional compliance-based safety. It is a tool for compliance safety experts and consultants. Why is that a problem? Because compliance-based safety has

---

* http://ehstoday.com/safety/ehs_imp_75429

always been a top-down directive activity that failed to engage the employees in the definition and management of the safety processes within a business. So when a management team or the resident safety expert decides to implement BBS, it is just another top-down decision forced upon others. Managing safety in this way, from the top down, rarely gets to the cultural issues that need attention.

Focusing on the process and not the person is a basic tenant of Lean thinking and, might I add, Lean Safety thinking. BBS polices people and we all know what happens when the police go away. Positioning a state trooper on the side of the expressway causes all cars to slow to the speed limit and a half mile down the road everyone is speeding up—policing works when the police are present. BBS promotes an ongoing culture of policing. Another problem is that this ongoing policing only focuses on compliance-related safety. The assumption is that if people follow the rules (OSHA), there will be no injuries. While observing workers being compliant or noncompliant, a BBS professional compliments the good behavior and points out the unsafe acts. If done often enough by a mass of people, this BBS activity will reduce injury incident rates. However, this approach fails to get to the real root causes—the underlying safety culture issues present in the business like "productivity is the number one focus of the business and has always been number one." In this environment, watching someone during a BBS audit and pointing out that he or she is not wearing the appropriate PPE does nothing to fix the deep-seated historical management problem of focusing on the bottom line and productivity ahead of safety. In addition, it appears that during a BBS audit you are doing it "to them" rather than "with them." This violates the Lean concept of employee engagement used to build trust.

In contrast, instead of focusing on just the unsafe behavior of the person performing a work task, a Lean thinker quickly digs into his or her Lean tool box and uses the "ask why 5 times" tool to both engage the employee and understand the underlying reasons for this disregard for compliance safety. They are going to dig down to the root cause cultural problems that exist. BBS should be retitled MBBS. That is management behavior-based safety and its focus should be the safety culture management has created via its actions and how to redirect that culture by engaging the workforce in proactive continuous improvement safety. Why? Safety cultures change when people make their own decisions to change. Watching them work and critiquing their actions does little to encourage them to think. Engaging a workforce (asking them to think) in safety improvement activities is a trust building activity that allows them to change their own minds. I am passionate about that Lean Safety approach. This is what Safety Gemba Walks accomplish.

The only time I was closely exposed to BBS practitioners was via a Ukrainian company that invited me to present at a safety conference it had organized. Its situation was unique in that it was quickly growing the business via acquisitions. The company's core competencies were steel production and mining, and it was purchasing old Soviet era mills and mines. The problem for the safety professionals was that at many of these sites there was little if any focus given to the safety of the workforce. The old Soviet era mentality that "life is hard and then you die" combined with top-down directive leaders in charge defined the safety culture in these facilities. Driving this safety culture change was the new owner's desire to make this Ukrainian company competitive with its counterparts in Western Europe. He was keenly aware that future customers would use safety as a metric when selecting their vendors. This was a wonderful opportunity for me to make a safety difference.

My contact, with whom I worked for months to plan and arrange my visit, was a Russian-born safety professional who is passionate about making a safety difference in a country and industries where safety was not always a priority. For example, the fatality rate (fatalities per 100,000 workers) in the Ukraine is around twice that of the United States. She and her safety co-workers were former "DuPonters"—individuals trained in the DuPont BBS philosophy. It was explained to me that the DuPont methodology requires you to work with your people to get to zero injuries because it is mostly people who make mistakes leading to injuries. People remove barricading, fail to use lockout procedures, take shortcuts, etc. This may be a sound approach to safety management and yet I feel this methodology gives focus to safety from only one perspective—the worker's activities relative to safety compliance. At first, the DuPont method seemed at conflict with my "the process is the problem—not the person" Lean thinking approach because I believe in eliminating the hazard when possible and thus eliminating the training and auditing required to help the person avoid the hazard. Of course, there is not one right approach—only different approaches that all have some merit. For instance, both my Lean Safety methodology and the DuPont method focus on the employees. My mission, however, was to help the conference attendees understand the Lean continuous improvement employee engagement approach to proactive safety improvement. What is different and dare I say new about my approach is that I ask people only to watch other people work and make their job safer and easier. It is not a focus on adherence to compliance requirements—it is a focus on the continuous improvement of the safety of an individual's work activities.

During my visit, I was given the opportunity to tour two of the company's production facilities. Prior to the start of the conference, I was given a tour of part of the Ilyich (Lenin) Iron and Steel Works. This 80-year-old facility has quite a history. For instance, two of its blast furnaces were disassembled and moved to Northern Siberia during WWII to prevent them from being destroyed by the invading German army. Metinvest purchased this facility just 2 years ago and has only started to make changes. Today, this drab aging site employs approximately 35,000 people! I was taken on a tour of the steel making and slab casting facilities before making a second stop to tour a plate mill that produced steel plates for shipbuilding. The manager of the furnace and casting area was waiting alongside the road when our minivan approached and stopped. He was an individual who was in charge and wanted you to know he was in charge. Using my contact as a translator, he explained to me that his hot mill was world class and better than most. He noted he had visited mills in other countries and his was indeed the best. He was not very interested in who I was or why I was there. Then after listening to him for about 20 minutes we were about to start the mill tour. As we departed the training room where we had been meeting, he pointed out a mirror by the door. You have all seen them. On it said, "You are looking at the person most responsible for your safety." Of course, I couldn't read it because it was in Russian. Then, this very serious old school manager reached toward the back of the mirror and pulled down a window shade that quickly covered the mirror. On it was a picture of Stalin, the former Soviet leader who was responsible for executing hundreds of thousands of people and imprisoning millions. He looked at me, grinned, and laughed. Since this plant has been in operation since the early 1930s, I bet this Stalin shade has been used to reinforce the understanding that you are indeed better off taking ownership of your safety rather than relinquishing it to Stalin! We then walked to the blast furnace/steel casting area. He guided me up onto overhead walkways so we could view the work from above. Since 18 years of my work history were spent at U.S. Steel Corporation, I was familiar with the processes I viewed. No evidence of Lean or continuous improvement activities was visible during this stop or my next at the plate mill. Leadership seemed stuck in the top-down directive Soviet past. For instance, as we walked on a dimly lit catwalk at least 50 feet above the plate mill, I observed some millwrights working on mill drive shafts down below. To describe their work area as dark would be generous. I suggested through my interpreter that they could improve the lighting in this maintenance repair area and it would certainly be better for the workers and improve their productivity. The response was that the lighting in the area "met the standard." This in no way reflects badly on the current managers because they (like all of us) are a reflection of their culture and working environment. Change is indeed the only constant, and we all have to face a changing world both at work and in our daily lives. I didn't know it at this point, but change was coming their way.

On the afternoon of the second day of the conference, I joined a bus full of conference attendees to tour the Khartsizsk Pipe Plant—a facility purchased in 2006. This site manufactures large diameter pipe for the oil and gas industry. The contrast between this plant and the Lenin Iron and Steel Works plant was stunning. I witnessed a completely different work culture and work environment due to both the application of the 5S Lean philosophy and the work of the internal safety professionals who conducted behavioral-based safety audits. The proud and Lean savvy leaders at this plant talked about how it was when they first arrived. Windows at the roof level were all missing and workers would stand around 55-gallon drums of burning wood to keep warm. They and their safety counterparts transformed a facility that was filthy and disorganized into a bright, well-lit, clean, and organized plant they were proud to show us or any visitors. It was a great example of what determined, focused operational leaders, working hand in hand with their safety professionals, could accomplish. While touring, I heard that the leader of this plant had just been transferred to the Lenin Iron and Steel Works, which I had visited three days prior. Since organizations take on the personality of their leaders, the culture at the Lenin Iron and Steel Works was about to undergo a big change.

It was noteworthy that the company's safety professionals, using their DuPont BBS practices, helped the management team of the pipe plant turn it into a showplace. This group of passionate safety professionals taught me that when no safety culture exists, using BBS as a foundation-building tool to help supervisors and workers understand what are safe versus unsafe actions is a worthy and beneficial compliance safety effort. They also reinforced my belief that the safety culture of a business is a direct, in the mirror, reflection of management. Management, along with the supporting safety professionals at this pipe plant, must like what they see in the mirror.

So, is BBS good or bad? I will leave that up to the compliance safety professionals to debate and decide. What I do know is that consultants cannot change the culture of a business. Management teams that hire BBS consulting firms to fix their broken safety culture are wasting their money because they must change their own behaviors to change their safety culture. No one can accomplish that for them. Another shortcoming of BBS is that it doesn't go far enough by requiring a review of the whole management system and its relationship to and impact on safety. For a culture change to be successful, you must engage the hearts and minds of your employees, and policing human beings, any human being, fails to win their hearts and minds. BBS at its core is still looking for "man failures" and fails to earn the trust of the workers being observed. If you don't believe me, just go to the Gemba and ask them.

Clearly, my intent was not to offend those who practice or consult in BBS and are the real experts in that safety methodology. I only want to highlight the limitations of this approach and challenge everyone to go beyond compliance. I think, in general, the BBS community of consultants already recognizes the limitations because they are all repacking their material and adding buzzwords like "culture," "engagement," and "efficiency." It should be obvious to them that in this "Lean" world that is built upon building trust, to promote a program that blames people (the "man-failure" approach) is a terribly flawed approach. So rather than just polishing up their marketing materials to sell the same product, I hope they are actually changing their products and their approach. My goal is to help everyone understand the difference between observing people at work in order to identify unsafe work actions (Lean Safety) and observing people at work to identify unsafe behaviors (BBS). My approach builds trust—their approach kills trust.

Chapter 4

Living Injury-Free Every Day versus Living Painkiller-Free Every Day

I had the opportunity to address a group of site leaders for a very large construction management company. I was impressed with a safety acronym they had developed and displayed. It was L.I.F.E.® (Living Injury-Free Every day).* Based on my passion for making work not only safer bur easier, I immediately created my own acronym—L.P.F.E. (Living Painkiller-Free Every day). The two acronyms are a great example that highlights the difference between compliance-based safety and the continuous improvement of safety. Living injury-free implies that if you follow the compliance rules and regulations, you will go home injury-free. This is a very worthwhile and meaningful compliance safety objective. Living painkiller-free implies that your work has been made easier so that you do not have to take painkillers every day, before or after work, to ease the muscle and joint pain. That is a continuous improvement safety objective.

The living injury-free objective is tied rather tightly to the common metrics of compliance-based safety. Measurements related to injuries, lost time, and near misses are the most common safety metrics found in every facility I have visited. They are both after the fact, lagging indicators of safety performance. I call them "wishing and hoping metrics" because managers and safety professionals sit around wishing and hoping they do not have any occurrences. The reason these are the established metrics for compliance-based safety is that OSHA requires businesses to track recordable injuries. They are both the cultural metrics of the safety community and a common language understood by all. What additional metrics would be displayed in a facility that also focused on the continuous improvement of safety? Safety improvements would be my first choice because everyone in the business can participate in both the proactive search and identification of ways to improve safety or the elimination of hazards that might cause injuries. How do you get started? Take a Safety Gemba Walk.

As part of a two-day Lean Safety workshop in Birmingham, England, the attendees visited a small manufacturing site on the second day. The Lean Safety workshop was one of the offerings at a Lean conference sponsored by a manufacturing magazine. Prior to the start of the conference, I, along with the conference organizer, visited the manufacturing site that the workshop attendees would visit on Day Two of my workshop. We wanted to ensure that the site operation's manager was comfortable with and prepared for the visit by the 20 or so workshop attendees. After exchanging greetings, I asked him why he volunteered to open his doors to the group of visitors. He looked at me and clearly stated, "Because my boss told me to!" Once I heard that he had been "voluntold" rather than him volunteering, I knew I had to earn his trust for the event to be successful. I quickly asked him

---

* http://www.turnerconstruction.com/about-us/safety

what he knew about Lean Safety, or the continuous improvement of safety, approach. Not much, it seemed. I then gave him an overview of the topic and my approach to engaging people in the continuous improvement of safety after which I asked him if he was planning to attend Day One of the workshop at the conference hotel. His initial response was no, but I could tell I had stirred his interest. Before leaving, he gave us a quick tour of his facility during which I had the opportunity to continue to build our relationship and earn his trust. Before we departed, he stated, "You know, I think I will attend the first day of the workshop." "Perfect," I thought! If I am going to guide 20+ people on a Safety Gemba Walk in anyone's facility, I need them to be on board with and understand the intent of the activity.

The initial defensiveness I experienced from this operations director is commonplace. Both business leaders and safety professionals are not initially open to having outsiders come into their house and look at safety. Why? I think it is legacy thinking tied to compliance-based safety. What they fear is someone telling them they are out of compliance because when that happens what they really hear is "You have done a poor job of managing safety here." No one wants to look bad or be embarrassed in front of others and that is the last thing I want to happen. I often say without trust Lean is a bust, and that is just as true when undertaking a safety culture change. Change is built upon a foundation of trust and I have to earn the leaders' trust before walking people through their facility. I often accomplish that by taking them on a Safety Gemba Walk first. They quickly understand we are not there to look at compliance safety and then, with eyes wide open, they are engaged and willing hosts.

So now let's get back to the manufacturing site and the second day of the workshop. The operations director did indeed attend Day One of the workshop and therefore had met and talked with the other workshop attendees. He greeted them warmly when they arrived at his site on Day Two and delivered a company and product overview. The site's core competencies were machining and mechanical assembly operations. Before we headed to the Gemba for an initial tour of the facility, he genuinely asked everyone to help him improve safety by identifying as many improvement opportunities as they could. We were given the required personal protective equipment (PPE) and then the operations director and an associate gave a tour of the facility to familiarize everyone with the site layout and the work processes. As we walked, I stopped at various work centers and would ask the group questions to help them recognize the opportunities for improvement.

Opportunities are everywhere in every facility around the world. The reason they exist is that the managers, supervisors, and safety professionals walk right past the workers working and only see people doing what they are being paid to do—work. But if they stop and really watch, with intent, the activities that take place to perform the work, they can uncover a wealth of opportunities to make work safer and easier. Taichi Ono, one of the creators of the Toyota Production System (TPS), developed a technique to help new managers learn to see the waste in processes. While giving a tour of a facility, it has been reported that he would stop and draw a chalk circle on the floor. He would then ask the new manager to stand in the circle and just observe what was happening in the work area. He would depart, leaving them alone, and would not return for some time. Upon his return, he would ask questions to determine if the manager had uncovered waste in the work processes. I do not draw chalk circles on the floor when I lead Safety Gemba Walks, but we do make frequent stops. Our objective is not to identify the seven deadly wastes (transportation, inventory, motion, waiting, over production, over processing, and defects), but only to watch people at work. For instance, if you were watching the popular game show "Wheel of Fortune" and as you observe one of the contestants bend over the podium at which they stand to reach out and down to spin the wheel, what do you really see? You might only see exactly what I have just described or like me, you might see individuals putting themselves into a position (back out of neutral) that might cause a serious back injury. While on a Safety Gemba Walk, participants are given some very basic guidelines they can use to find opportunities to make work safer and easier. They relate to the identification of risks that can cause soft tissue injuries.

What do they look for? It really is a simple concept with simple instructions, and yet repeatedly I am surprised at the reaction of people on Gemba Walks. They often have their "ah-ha" moment when I ask questions like, "What did you just see there?" It could have been someone bending over to measure a part in a gauge. Managers, supervisors, and safety professionals walk by people bending over to complete work tasks all day long and only see people working. I see an injury risk. While we continue to observe, I ask them what could be done to improve that situation—to keep the individual's back in a neutral upright position. They quickly identify multiple opportunities like elevating the gauge or positioning a stool in front of the gauge that would allow the operator to sit/stand with his or her back in neutral while gauging the part. As we continue to walk and observe, I continue to question them and they continue to learn to see the inherent risks present

in everyday work. The four guidelines they and anyone can utilize to identify workplace risks are as follows.

- **Body parts out of neutral**—Look for any body part out of neutral.
An out of neutral example is a person bending forward as little as a
half inch. Working with your back out of neutral results in a sore back.
Having to reach up to get something routinely means a shoulder is out
of neutral. When a person has to twist to grab material to his or her left
or right with his or her feet stationary then the torso is out of neutral.
Awkward wrist movement that takes your wrist out of neutral can lead
to carpal tunnel syndrome—a medical term none of us had heard prior
to 15 years ago!

  To help you better understand this core concept of Lean Safety, here
is an example with which you can all identify. Just imagine yourself
boarding one of Boeing's new Dreamliner airplanes. Shortly after taking
off on this technological marvel of the 21st century, you will be con-
fronted, or maybe rammed, with technology from 1955.

  I am talking about the drink cart that is rolled, pushed, and shoved
down the main aisle of many planes by the flight attendants. I recently
flew home on a flight from Dallas. This flight, along with every other
flight, allows for plenty of time to practice the Lean Safety skill of
"observing a work process with an eye for safety improvement." The
flight attendants I observed had to constantly twist and contort their
backs and necks out of neutral while leaning over in an attempt to
locate the correct drink can in the trays of cans stored under the cart.
They repeatedly slid out trays that were unstable and hard to slide in
and out because they only lay on guide rails. If they pulled the tray out
too far, it and the contents could fall to the floor. Therefore, they are
required to hold on to the front of the tray while bent over, in low light
conditions, searching for that elusive last can of apple juice. Once the
passengers in the immediate area were served their drinks, they now
had to move this ancient cart down the aisle. Next, the flight attendant
repeatedly stabbed the area around the cart wheels with her foot in an
attempt to hit the lever that would release the cart wheel brake. Then
with the help of another flight attendant, they pushed and shoved the
cart forward. Well, not really forward—it was more of a zigzag route
caused by the wheels, which due to a lack of maintenance, did not
roll very well. Just imagine this vintage cart, which deserves a spot in
the Smithsonian Air Museum alongside the Wright brothers first plane,

scrapping the new surfaces off the arm rests on a new Dreamliner. More importantly, imagine the effect of the contorted and stressful actions on the bodies and limbs of flight attendants. A goal of Lean Safety is to keep an individual's body parts in a neutral position while working. Observing this work task with an eye for improvement would provide a lengthy list of opportunities for improvement.

■ **Straining**—Straining to turn or move anything can cause soft tissue injuries. An example I use in my workshops is opening a pickle jar. Speaking from experience, as we age, opening a pickle jar does not get any easier unless you have a Jar Pop (http://www.wdrake.com/buy-jar-pop-opener-305021). This simple plastic device has a hook that rests under the lid while the handle extends horizontally from the center of the lid to beyond the lid where you hold it. As you give the handle a slight lift, you hear a "pop," which releases the vacuum-sealed lid that now easily spins off the jar. So, if you observe someone straining to complete a work task, you are obliged to find a version of a Jar Pop that will eliminate the risk.

■ **Lifting**—If someone is observed lifting anything that appears to be heavy, a solution may help prevent an injury. I am amazed when I visit some facilities and observe the amount of manual lifting that still occurs. Many facilities have a limit on the weight someone can lift—for instance, 50 pounds. It is unrealistic to set these types of safety rules because individuals could weigh from 90 to 250 pounds and differ in height by 18 inches. One-size-fits-all rules like lifting restrictions just don't work. Near the end of my workshop, I often use a PowerPoint slide that has a Pareto chart showing the types and frequencies of injuries in a calendar year. The tallest bar is for back injuries. I then ask each table to develop a plan to reduce the frequency of back injuries in the upcoming year. Once when the workshop attendees were almost all safety professionals, more than one table had lifting training as one of their actionable items. Upon hearing this, I asked for a show of hands if they wanted to lift for a living. No one raised his or her hand. So, why do we ask people to lift when there are so many lift assist devices, like vacuum lifts, that can be purchased and put into place?

■ **Repetitive work tasks**—Our understanding of ergonomics and how you can apply this scientific methodology in work settings to reduce soft tissue injury risks increased dramatically starting about 15 year ago. This is when OSHA was going to institute an ergonomics standard to which all businesses would have to comply. It is also the time when we all became aware of the words carpal tunnel and their meaning. In the end, the idea of a common, one-size-fits-all ergonomic standard was abandoned by OSHA in favor of a more targeted approach that would look at specific industries. Despite the shift in awareness of the risks, injuries and cost of soft tissue injuries in the medical, insurance, and legal communities drove changes in workplaces. Despite all of those efforts, repetitive work still exists and when it is observed on a Safety Gemba Walk, it is a target for improvement.

After the general tour of this UK site was completed, the attendees were split into four smaller groups. They were given free rein to go anywhere on the plant floor with the objective of engaging the workforce in discussions that would help to identify as many potential opportunities for improvement as they could in 1.5 hours. When their time expired, we all gathered in a training room for a report out. I always ask for a volunteer from the site staff to record the opportunities on a flip chart or white board. Having someone from the site record the opportunities means he or she will ask the right questions to ensure he or she understands the opportunity, thus improving the odds that the task will be implemented. To my surprise, the site operation's manager offered to be the recorder. The same person who explained he had been "voluntold" to host the second day of the workshop just two days earlier now suffered hand cramps as he recorded approximately 100 improvement ideas. Conducting routine Safety Gemba Walks going forward may prevent the risk of soft tissue injuries to his hands because the list of ideas generated should be smaller! This event was interesting to me because it is a great example of how someone who is at first defensive about others coming into a facility to look at safety in the end can find so much value in the exercise. Before we departed for the day, he sincerely thanked everyone who attended and promised them that he would work to implement all of the opportunities they had identified. So if a group of outsiders can identify 100 safety improvements in just a few hours, imagine what all of the employees in a facility can do to help each other L.P.F.E.. All it requires is training so everyone can learn to see the opportunities.

Chapter 5

A Safety Walk versus a Safety Gemba Walk

Safety walks have an agenda (probably a standard form), a targeted location or route, and people trained in compliance safety who conduct the safety walk. The sole focus of traditional safety walks is to identify noncompliance situations. To find those noncompliance items, the participants on a compliance safety walk usually only look at "things" and "stuff," not people. A Safety Gemba Walk only has a targeted location or locations and a skilled facilitator who, using his or her interpersonal skills, engages workers in a dialogue in order to identify opportunities to make their work safer and easier.

When I was an operations manager, I went on traditional compliance-based safety walks. Usually a manager, like me, was paired with a shop floor associate who was a member of our employee-based safety committee. We used a standard form and a predetermined route to search for noncompliance items. We would record our findings and often talked with the employees in the area and engaged them in immediately correcting the compliance problems we had discovered. For instance, if on a pedestal grinder we found that the front edge of the steady rest next to the grinding wheel was too far the from the wheel, and therefore creating a gap wider than OSHA allows, we would ask a machine operator in the area to get the required tools and adjust the steady rest forward to narrow the gap. We corrected a compliance problem, or some would say a violation, but we did nothing to ensure ongoing compliance with the OSHA requirement. Our safety walks would uncover numerous violations just like the one described and almost every one of them had to do with "something" out of compliance. By reading the information on the following safety walk form that I pulled off the Internet, it will be clear that the sole intent of compliance safety walks is to ensure compliance.[*]

Navosh Safety Walk-Through Checklist		
References: (a) OPNAVINST 5100.23 series (b) OPNAVINST 5100.19 series (c) Code of Federal Regulations (29 CFR) (d) National Fire Protection Association Codes		
Walking/Working Surfaces:	*Yes*	*No*
Work center floors clean and dry		
Hangar decks clear of FOD and spills		
Unprotected/unidentified trip hazards		

---

[*] http://www.public.navy.mil/comnavsafecen/documents/aviation/maintenance/navosh_wlkth_org.doc

Stairs safe (secure rails, treads)		
Maintenance ladders in good repair (rungs, feet)		
Scaffolding/maintenance stands >5 ft have top and mid rails, toe-boards, wheel locks		
Electrical:	Yes	No
All disconnects and circuit breakers labeled		
36″ clearance around circuit breaker panels (no obstructions)		
No exposed live wires or circuits		
Circuit breaker panels/receptacles have no holes, open slots or removed knockouts		
All receptacles, switches, and boxes have covers in place and in good condition		
No permanent extension cords used in place of fixed wiring		
Explosive-proof lighting/machinery in use where needed		
GFCI circuits used in wet areas		
Plug ends do not have ground pins removed		
Machinery Guarding:	Yes	No
Barrier guards on moving machinery parts, belts, and pulleys		
Point-of-operation and pinch points guarded and marked		
Fan blades guarded (<1/2″ opening)		
Fixed machinery anchored to deck/work bench to prevent movement		
Bench grinders (1/8″ tool rest and 1/4″ tongue adjusted from wheel)		
Band saw blades guarded above guide rollers		
Safety zones around shop equipment		
Abrasive wheels in good condition with no evidence of side grinding or nonferrous materials being ground		
On/Off/Kill switch easily accessible		
Hazardous Material:	Yes	No
Only approved lockers in use (3-point locking mechanisms and bungs installed)		

(continued)

HAZMAT lockers properly identified (Flammables, Oxidizers, and Corrosives)		
HAZMAT labeled with approved labels. No NFPA labels (diamond shaped)		
Lockers contain only material listed on AUL, Inventory, and have MSDS		
No Smoking signs posted and observed		
Are MSDS readily available and accessible (no physical or administrative barriers)		
Rooms used to store flammables properly ventilated and use explosive-proof equipment/lighting		
Are refrigerators properly labeled as to contents (either "Food Only" or "HAZMAT-No Food")		
Are dispensing containers (55-gal drums) grounded to prevent static discharge		
General Safety	*Yes*	*No*
Plumbed eyewash stations/showers flushed weekly (verify inspection record)		
Portable eyewash stations purged monthly (verify inspection record)		
Safety zones painted around eyewash stations (suggest green) and kept clear		
Firefighting equipment inspected monthly, clear of obstacles, and has painted red zone		
PPE in good condition (elastic, lenses material integrity)		
Weight Handling Equipment (WHE) inspected and load tested		
Space heaters have approved auto-tip over protection		
Coffee makers approved by Fire Marshal and on metal surface/tray		
Industrial Hygienist Survey posted in work center		
DOD/NAVOSH placard and policy statement posted (current Commanding Officer)		
Unstable shelving secured to prevent tipping hazard		
Are members smoking in designated areas only		
Are applicable warning signs posted (hearing, sight, foot)		

Safety Gemba Walks, in contrast, do not focus on "things" but rather people. The following example from a recent vacation will help you understand what I mean. Lean thinkers like me cannot turn off the "Lean thinking" portion of their brain while on vacation. While walking from the Notre Dame Cathedral back to our hotel in the 7th arrondissement located on the left bank of Paris, my wife and I decided to take a break. We selected a park bench on the bank of the Seine River where we could bask in the glorious sunshine and count our good fortune. We had spent the prior ten days in Ireland where cool rainy weather had been as much the norm as driving on the left side of the road. Then on a whim, without any existing plans, we walked into a travel agency in Galway and made plans to fly to Paris the next morning to finish our vacation. It was completely spontaneous, which made our five days in Paris even more magical.

While sitting there looking north across the Seine, we could see the massive Louvre museum to our right and the 3300-year-old Egyptian obelisk sitting in the Place de la Concorde, its gold top reflecting the mid-day sun, to our left. Then out of the blue Parisian sky, my Lean Safety antenna was signaled to observe and improve a work process. Almost directly in front of me was a city worker bagging grass clippings. He, like me, was not a young man. His back, which had a gentle slope forward, symbolized a lifetime of work. The grass clippings were piled in a graveled area some distance from the actual lawn that had been cut. He was surrounded by four clear plastic bags that he had already filled and was just starting a fifth. The process he was using certainly contributed to or was responsible for his back curvature, depending on how long he had performed this job responsibility. He was using a rake but it differed from the rakes I have used. This four-tined rake looked as if it started as a pitchfork until someone bent all four tines 90 degrees from the handle. While holding the handle parallel to his body, he slid the tines into the grass pile. Next, he stepped with his left foot onto the grass clipping now on the rake tines to compact them before continuing. Then, while holding a plastic bag in his left hand and the rake in his right, he raised his right arm, taking his right shoulder out of the neutral position, to align the grass clipping on the rake tines with the opening in the plastic bag. He then attempted to insert the clippings into the bag. Some made it in and others fluttered back to the pile. He was looking directly down, with his back and neck bent, during the entire operation. Had my command of the French language allowed me to do more than order outstanding food and great wine, I would have engaged him in a process improvement discussion intended to improve the safety of the work he was performing.

My eyes were now glazed, I had forgotten about the glorious "City of Lights" in front of me, and I was asking myself, because I couldn't communicate with the man, some simple questions. How did the clippings get from the lawn to their current location? Didn't they know they make mowers with bagging attachments? Or, did they dump the clippings from the mower attachment here so they could rebag them? I had to admit that this was indeed a possibility because after all this was a government worker in a country and city better known for worker strikes than worker ingenuity and productivity, so I shifted my thinking from correcting the root cause to just improving his safety.

When we first arrived in Paris, I noticed there were few conventional trashcans. Instead, they had metal hoops welded to upright posts that were anchored to the ground. The top of a plastic trash bag was slipped over the hoop and a rubber bungee cord type device was used to secure the bag to the hoop. To remove the bag you simply loosened the rubber cord. I was impressed with this creative method of trash collection that eliminated dirty, smelly trashcans and the requirement to lift them to empty out the contents. Then I noticed just to the left of our bench one of these trash collection stations. I wanted to immediately move some grass clippings next to it, install his plastic trash bag onto the hoop, straighten the tines on his rake, and fill the bag while maintaining a more upright back and head position with my shoulders never getting out of the neutral position.

Critical to understanding Lean Safety is to accept that to understand the work being performed and the risks inherent in the task you must observe the person at work. Previously I described correcting an out of compliance tool rest on a pedestal grinder during a compliance safety walk. On a Safety Gemba Walk, one would observe a person grinding a part on the pedestal grinder as opposed to looking at the pedestal grinder to ensure it is in compliance. Therefore, there is no comprehensive Lean Safety Gemba Walk form. You simply walk through a facility and observe people at work while engaging them in discussions about making their work tasks safer and easier. Therefore, there is no form like the one used for compliance safety walks. The only documents or forms available are used to record the opportunities as they are identified. On them, all of the potential opportunities to make work safer and easier are recorded. The next pages are examples of two forms that I use when guiding individuals on a Safety Gemba Walk. They are usually printed on one sheet of paper so that the opportunities can be recorded on one side and on the other side they have a chance to compliment the site by listing the positive things they observed during the Safety

---

LEAN SAFETY GEMBA WALK OBSERVATION FORM

Date: _____

Observer: _____

WHAT DID YOU OBSERVE THE INDIVIDUAL DOING?

CATEGORIES
1. Material Handling
2. Material Storage
3. Workstation layout
4. Ergonomics
5. Material Flow
6. Other

_____

_____

_____

_____

_____

_____

_____

_____

_____

_____

_____

_____

_____

_____

_____

_____

OPPORTUNITY

Gemba Walk. Each Safety Gemba Walker has at least one form and I set the expectation that they will fill it up before we complete our walk.

When I tour around a facility for the first time, I look for some things that all Lean thinkers might observe. First are the seven deadly wastes and their sources. If you understand and can identify the seven deadly wastes, then you are well on your way to becoming a Lean thinker. This understanding is the foundation of all continuous improvement programs—CI, total quality, organizational excellence, Six Sigma, Lean, etc. Those seven wastes are:

1. **Transportation**—The unnecessary movement of products, paper, information, etc. adds no value, only cost.

LEAN SAFETY GEMBA WALK OBSERVATION FORM

Date: _____

Observer: _____

**WHAT DID YOU OBSERVE THE
INDIVIDUAL DOING?**

_____

_____

_____

_____

_____

_____

_____

_____

_____

_____

_____

_____

_____

_____

CATEGORIES
1. Material Handling
2. Material Storage
3. Workstation layout
4. Ergonomics
5. Material Flow
6. Other

WELL DONE!

---

2. **Inventory**—Inventory masks many of the problems in a business. It is like an insurance policy. Equipment breaks—it is okay we have inventory. Suppliers deliver late—it is okay we have inventory. Employees report off—it is okay we have inventory. Lowering inventory levels forces a company to address all of these and other problems.

3. **Motion**—Steps in a business process that take time but add no value to your services or products.

4. **Waiting**—Inactivity caused by a lack of material or information.

5. **Over-production**—Making products in excess or before your customer requires them.

6. **Over-processing**—Adds cost because of extra steps or delays in a business process.
7. **Defects**—Services or products that do not conform to your customer's expectations.

In addition to looking for the seven wastes, a Lean thinker would also give focus to product flow, material handling methods, storage containers used, and the plant layout (including the overall department and work cell layouts) while touring a facility. I am not your normal Lean thinker because as I observe these things, I am thinking about their impact on the individuals who work there. With that in mind, let's take a theoretical Lean Safety Gemba Walk. As we do, we will be observing the following four common themes present in any facility.

Product Flow

When I tour around a plant for the first time, I always ask about and want to understand the product flow. I want to walk the flow starting with raw material delivery and ending with the shipment of finished goods. Product flow can be defined at the plant, department, and work cell level. The product can be physical parts, information, or in the case of a hospital, real people. A Lean thinker does this because improving flow will reduce the delivery cycle time to customers and that is the goal of Lean. Every time the flow stops and parts, information, or people are stalled, I ask why. Asking why causes people to reflect on and truly begin to understand their operation. My role as a consultant is to challenge people to dig deeper and understand the opportunities for improvement that exist. From a Lean Safety perspective, poor material flows, with starts and stops, require additional material handling. Whenever material is handled, people are in harm's way. Therefore, improving the flow of the work through a facility is one of the first critical steps on the Lean journey, and it is also a great way to minimize injury risk. Lean improvement and injury risk reduction go hand in hand.

Material Handling Methods

As I walk the flow of a facility, I closely observe the material handling methods. It doesn't matter if you are using a hoist, a forklift, or your back—material handling of any kind is an injury risk. Therefore, minimizing material handling reduces the risk of injuries related to material handling. Often, reducing the amount of material handling is accomplished by limiting the amount of time a person spends on a particular task. Large package handling facilities, like UPS or FedEx, where they load and unload semi-trailers by hand, might utilize shifts that are only 4 to 5 hours in length. These modern facilities have high-speed conveyor systems that read carton bar-code labels, and direct and redirect cartons with diverters to the correct truck loading dock. However, at the dock someone still has to stack the cartons in the truck in order to best utilize the available space. It is like working on a giant jigsaw puzzle. On a smaller scale, in almost all facilities and especially in large distribution centers, individuals have to arrange boxes on pallets and the pallets are usually on the floor. Getting pallets off the floor is an easy opportunity to make work safer and easier. In the end, the goal is to eliminate lifting whenever possible and if this can be accomplished the risk of injury is reduced as is the cycle time of the work process.

Storage Containers

While walking the flow in a facility, I am constantly looking at all of the different storage containers and how the associates interact with them. Large deep containers like wire baskets and steel hoppers store a lot of material but require people to bend over to put parts in and take them out. These large containers require a forklift to move them and are often legacy containers linked to the MRP big batch production mentality. Today, Lean thinkers reduce batch sizes on their journey toward Lean production nirvana—one-piece flow. Most manufacturers cannot get to the Subway sandwich shop model of building and delivering one at a time, but they can certainly reduce their batch sizes. The other sources of many of the containers that fill our facilities are our suppliers. Most often, the supplier has defined the box size, which in turn defines the weight of the box, rather than the customer. Material is delivered in large unwieldy boxes with the parts disoriented inside. Most people pay no attention to the associates who now have to handle the boxes and remove the parts from the boxes. Investigating how material is stored and delivered leads to a gold mine of injury risk reduction opportunities. While on a Lean Safety Gemba Walk with an industrial distributor, I observed an individual cutting a clear plastic wrap from a large diameter hose. The hose, which was about 20 feet long, was lying on the floor. The associate was bent over the hose and used a box cutter to slit the protective plastic cover as he crab crawled across the floor. I asked, "Why is the hose covered in plastic?" The obvious answer was to protect the hose. I then asked how the hose was shipped to the facility. In a corrugated box was the answer. So, is the plastic cover required? If not, ask the supplier to stop putting it on so you can stop not only cutting it off but also paying someone to haul it away in your trash. The last opportunity related to containers is to look at how you ship your products to your customers. How are the containers you send to them affecting their cycle times? Are you minimizing their injury risks or are you adding to them?

Plant Layout

How a plant, department, work center, office, or any work area is laid out has a huge impact on the flow of products and in the end the delivery cycle time to a company's customers. A good Lean layout fulfills relationships, is dense, and results in good flow and therefore velocity of product to customers. When defining a new layout, you are forced to address the three previous items just discussed—flow, material handling methods, and storage containers. Therefore, starting with a blank piece of paper and some cutouts to represent your departments and equipment is a very worthwhile exercise if you are serious about getting Lean and serious about injury risk reduction. An industrial engineering methodology uses block and relationship diagrams to define new layouts. The same process can be used for an entire facility, a department, or an individual work center.

Block Diagramming

The first step is to identify all of the blocks. These could be departments if you are working at the plant level, or equipment or desks if you are at the department level. Each block represents an activity center required to run the business and serve the customer. As an example, for a manufacturing plant layout you would have blocks that represent each production area, finished goods, shipping docks, maintenance, receiving dock, office, etc. A department level layout would have blocks representing the individual pieces of equipment, tooling storage, component inventory, etc. The next step in defining the new layout would be to clarify the relationships between the blocks.

Relationship Diagramming

At a Subway sandwich shop, they have clearly defined the relationships between the ingredients used to build a customer's sandwich. The sandwich wrapper is first, followed by the bread, which is the base upon which they will add the other ingredients. By locating the sandwich wrapper next to the bread, they have fulfilled the relationship between the two. Next are the meat, extras like cheese, lettuce, tomato, etc., and finally the condiments. This simple example clearly helps people understand the benefits of a layout that supports the flow of product. Yet, on many Lean Safety Gemba Walks, I observe plants that have confusing and unclear paths for their products. Parts are routed in batches from department to department and sit in queues waiting to be processed. How can you change that? Take the block diagram that was completed and using lines to represent the strength of the relationships connect the blocks. A very strong relationship, like that of the sandwich wrapper and the bread at Subway, would require you to draw three lines to link the two blocks. Any two blocks that are connected by three lines would require you to move the department or equipment next to each other. Continue this process for all the blocks using three, two, or one line to connect all blocks. The more lines the stronger the relationship and the case for moving the blocks to fulfill the relationship. In the manufacturing plant where I worked for over 20 years, we used this process to redefine a new layout for our entire facility. It was a very worthwhile exercise at many levels. It greatly improved flow, reduced material handling injury risk, reduced internal costs, and challenged everyone's thinking. I would encourage you to complete block and relationship diagrams for your facility. It will be an eye opener.

The last and probably the most important skill required by a Lean Safety Gemba Walk facilitator is his or her ability to engage individuals in honest process improvement discussions. Here are some "rules for engagement" that will help anyone be successful.

Rules for Engagement

Engaged—Greatly interested or committed. Wouldn't every business want employees who are greatly interested in continual improvement and committed to customer service? Yes has to be the answer. If managers and supervisors want to engage everyone in the business continuous improvement process, they must change their behaviors first. Alternatively, managers and supervisors must change their behaviors for a culture change effort to succeed. Here is a short list of demonstrable actions that can be taken to support employee engagement.

1. Adult-to-adult conversations. For too many years, workplaces have been the home of parent-child relationships and conversations. This type of work culture is one where workers are given direction and are expected to ask their supervisor for approvals for everything. Today it is understood that when supervisors and managers tell people what to do, it removes the responsibility of the employees to solve the customer service problem that exist in the business. Today, managers and supervisors must ask the right questions, not have the right answers. Information for decision-making must be pushed down the organizational chart so that the least empowered can begin to make business decisions. Treating everyone as an adult and having only adult conversations will demonstrate to everyone that the work culture is beginning to change. This does not happen organically; therefore, training in coaching and communications skills may be required.

2. Eye contact. This simple but very effective technique lets everyone you talk with know you are having an adult-to-adult conversation. Looking anyone right in the eye gets his or her attention like nothing else.

3. Active listening. I believe active listening to be the most underutilized interpersonal skill we have at our disposal. In today's world, we are all constantly distracted by our smart phones. They are a self-contained microcosm of our life that contain our contacts, photos, tools, and access to information on any topic. As a volunteer who helps to organize the largest Lean conference in the world, I have watched a dramatic change in the interaction between people who attend. The conference organizers are always looking for ways to help people network at this event. Longer breaks are provided in between the value stream sessions to promote and support networking, yet when the breaks occur everyone streams out of the meeting rooms and into the hallways and they almost all do the same thing. They turn on their smart phones and retreat into their own private world rather than interact with others. The ability, or maybe the desire, to listen without interruption or distraction is a required skill set for a leader of change. To look someone in the eye and actively listen to what he or she is saying without formulating a response while the person is talking is one of the best gifts you can give another human being. If you don't believe me, just ask your spouse or partner.

4. Initial introduction. When initiating a discussion intended to engage someone, I always begin by introducing myself, and then I ask about them. How long have you worked here? How do you like working for this company? The next step is to inform them why I am there and what I hope the outcome of our conversation will be. For instance, "I am here to observe your work process in order to make work safer and easier. Since you are performing the work, I will be observing you. I hope you are okay with that because it is the only way I will be able to understand the work process. I am very interested in your ideas on how to improve the work you do. After all, you are the process expert."

5. Small talk. Thinking that you don't have time for small talk is a big mistake. To get to know someone, you have to find out about him or her as a person. What are their interests? What do you have in common with them that will provide the connection to begin trust building? At the start of each workshop, I ask the attendees to introduce themselves by providing their name, company, role at work, and one thing they are passionate about. Their passion is an entry point for conversation and a differentiator that helps me remember them as a unique individual.

6. Seeking approval. Mentioned previously but worth restating is the technique of seeking someone's approval to observe him or her at work. Equally important is the explanation of why you are asking and ensuring the person is clear about the expected outcome.

7. Ask for help. Always acknowledge their expertise and ask for their help in solving the problem at hand. People will rarely refuse a request for help if you make eye contact and ask sincerely.

If managers focus on and master the above techniques that allow them to truly engage their employees, they will be on their way to becoming people centric leaders—leaders who believe their employees are the most important element of the business. The statements "Our employees are our most important asset" and "Safety is our first priority" are hollow statements if the actions of the business's managers do not reflect a belief in the statements. The old adage "actions speak louder than words" is worth remembering when you are in a leadership position and asking others to change.

Now that I have set the stage by providing both some educational background material about what I, a Lean thinker with a passion for safety improvement, look for when I walk through a facility and how I interact with the people working there, let's journey together on some Lean Safety Gemba Walks.

CASE STUDIES

I would like to acknowledge all of the sites that contributed to this book simply by opening their doors and allowing me and others to first engage their workforce and then together focus on safety improvement in their facility. It takes a courageous management team to open their facility's doors to outsiders. Thinking criticism might be the only result of this action has prevented some sites from agreeing to host Lean Safety events. It is almost guaranteed the leaders at these sites were only thinking about OSHA compliance safety. This fear of looking "bad" has also caused companies to decline my offer to have their name included in this book and that is okay. I understand that having a company name in a book about safety improvement implies that the company needed safety improvement. I also clearly understand every company needs safety improvement—both compliance and continuous improvement safety improvement. Some managers do not want to admit that. Managers who truly understand Lean approach this touchy issue differently. They understand the benefit of having outsiders walk through their facility and actually look forward to the expected outcome, which is feedback that provides opportunities for continued organizational growth. They do not see the feedback as criticism but instead understand they are being provided with opportunities for improvement. They also understand that sharing what they have learned with others in books, presentations, and videos, even when it exposes some weaknesses, allows them to learn from others in return. Companies that belong to consortiums buy into this concept of shared learning because it is the basic building block and benefit of joining a consortium. A manager at one of the sites that agreed to have its company name appear in the book stated, "That is the way we were." This implies that he and his organization learned from the experience and he believes other organizations can now learn from them.

I was invited into each of the case study sites described in this book and I will be forever grateful to each of them. Each of these sites is helping me change how people view workplace safety by adding to this body of knowledge. Thank you to each site for having the courage to acknowledge the fact that safety can be a continuous improvement activity rather than just an activity related to compliance. Some sites approved my request while others declined to have their company name appear in this book. Regardless, each case study is based on a site visit and contains insights to help the reader better understand the concept of Lean Safety Gemba Walks. I hope you share my excitement as you discover new ways of making a safety difference while reading the case studies. Let's begin our walk.

Case Study 1

Metal Fabrication Operation, Ohio

An iPhone, an iPad, and a laptop computer are woven into my daily routine. I am rarely if ever without one of them in my possession from the time I wake until I call it a day. That changed briefly a few years ago when in late July I left the laptop and iPad at home and flew to a location in Northern Ontario devoid of cell towers—a place where my iPhone was as useless as a cassette in an iPod.

I was surprised at how easy it was to drop out of the digital age. Three fishing partners and I climbed into a De Havilland "Beaver" float plane and were flown 170 miles north of Nestor Falls, Ontario to Keeper Lake. Dropping out of the digital age was like a trip back in time—a time when people talked to each other and their thumbs were used for more than texting. We had the chance to laugh and joke like young boys enjoying summer days that seemed to last forever. Escaping from emails, Google searches, the latest world news, phone calls, and the constant checking of a smart phone was not a hardship but a blessing. I hope each of you plans and enjoys an escape from your digital devices in the near future.

I did not turn my phone back on until we crossed the border on our return trip in order to avoid the excessive roaming charges. As soon as I switched it on, a flood of emails and voice mail messages downloaded. One of them was from someone asking if I could visit his facility in Ohio on rather short notice. I called to discuss the request. What I found out was this company had formed a team composed of operational people from four of its sites to identify improvements to a common manufacturing process that was responsible for multiple injuries across its corporate sites. In the follow-up emails sent by my company contact to coordinate and plan my visit, he referred to the team as a jishuken team. I was not familiar with this new Lean term.

> So, what is *jishuken?* As stated on Toyota Motor Manufacturing Kentucky Inc.'s website, jishuken is a "management-driven kaizen activity where management members identify areas in need of continuous improvement and spread information through the organization to stimulate kaizen activity." In other words, while kaizen typically involves all employees on the shop floor, jishuken requires managers to conduct hands-on kaizen activity on the factory floor. This helps managers take ownership of kaizen and create a culture of observing problem areas in their natural state.*

This helped to clarify why the team was formed. Apparently, the injury rates associated with the work task were unacceptable, so management decided to give it focus. This is reactive safety and is part of a good compliance-based safety program. Using the data (injury rates by type and location) to drive action is a continuous improvement activity that will help to reduce or eliminate the occurrences. I only mention this because in a continuous improvement (Lean Safety) culture, the risks may have been proactively identified and solutions put in place before the injuries occurred. However, since that was not the case here, I agreed to schedule a visit and assist them.

I was invited to participate in the second meeting of their cross-facility jishuken team. Their first attempt at identifying improvements to a common metal fabrication work process, positioning large parts to be tack welded, had resulted in some improvements, and yet additional improvements were required. My predetermined goal was to help the team dig deeper and uncover additional opportunities to reduce the injury risks associated with this physically challenging work process. I had been prepped with the information that some on the team thought they were finished improving the work task and could not understand why they were being asked to take another look. Less than two weeks later, I was sitting with the team on our first of two days together. After introductions, I gave a brief overview of the Lean Safety philosophy. The team left the meeting room armed with the subtle but very important understanding of the difference between just

---

* http://www.industryweek.com/workforce/just-say-jishuken

watching a work process to find improvements and watching a person work to understand and improve the safety of the work process.

I believe the jishuken team's familiarity with the work task, using cranes and their physical strength to move and align heavy metal parts, had gotten in the way of seeing all the opportunities present. Asking internal people who are familiar with a work process to improve that process is not the best approach. In this case, even though they were from different facilities, each of those facilities had the same setup, tack weld, and weld operation present. Alternatively, forming a cross-functional team so you have some individuals unfamiliar with the work will most often lead to better results. The "external eyes" on cross-functional teams will ask the difficult questions that challenge the team to dig deeper. During this event, I filled the role of the outsider.

Together we proceeded to the Gemba where I was guided on a general plant tour before proceeding to the work center where we would together observe the work process that had caused so many back strains and injuries. As stated earlier, the team had already observed this work process both at this site and at their own facilities. They had defined some improvements but because they were too close to the process, they accepted many of the unsafe actions associated with the work process. For example, all of the flat parts that were to be welded into the final structure were delivered to the facility lying flat on pallets. The team and I watched as the welder repeatedly searched though the inventory to find the right components. He then leaned over with his back out of neutral to either manually lift the smaller parts or attach a plate clamp to lift the heavier parts with a hoist. To the jishuken team and possibly to all the managers, engineers, and safety professionals who walked through this factory, this was not seen as a problem because it is just the "work required to get the job done." They are all blinded by familiarity. They and operational people all over the world walk through their plants day after day, week after week, month after month, year after year, and fail to recognize the opportunities to make a difference is someone's life by making his or her job safer and easier. However, I rubbed their noses in this particular soft tissue injury risk activity and many others that I observed. My approach is to ask a series of questions until the Gemba walkers learn to see safety differently. Once they get it and start adding opportunities to the opportunity log, I know that I can back off and let them go. Regarding the constant bending to pick up or clamp parts for lifting, I asked them how could we keep the welders' backs in a neutral or upright position? Could the flat parts be stored in a vertical position? Could the smaller parts be delivered on a cart? Could the pallets be set on a fabricated steel table? I kept peppering them with questions until they began to discuss all of the possibilities for just this one process step.

We then continued to observe the current state fabrication processes of positioning the steel parts onto a large welding table and tack welding them together before finishing welding the part. A list of approximately 30 improvement ideas was generated. Lists are easy to build but eventually the hard work of testing all of the potential changes must begin. Therefore, it was time for me to extend trust, back off, and let the team take ownership of the improvements. I was only a facilitator responsible for guiding the jishuken team on their journey of discovery—learning to see safety differently. While they began to discuss the opportunities and their next steps, I left to take the site manager on a Lean Safety Gemba Walk.

During the planning discussions with my site contact prior to my visit to the site, we agreed I would attend and observe an accident investigation meeting. Accident investigation meetings are open books that clearly describe the safety culture in a business. Too often, the only focus of these meetings is the individual who was injured or nearly injured in a near miss occurrence. Very early on in my Lean Safety workshops, I ask for five volunteers, give them roles and scripts, and have them conduct an investigation meeting in front of the other attendees. I have scripted the skit so that the focus is on the injured individual with the intent to issue discipline because he or she failed to follow the safety rules regarding lock out tag out. This role-play skit mirrors what is happening all over the world. The focus is on the person, the injured, rather than the process. Why is that? Because compliance-based safety, which is the norm, is a top-down directive activity. Safety professionals are viewed as the safety police who are responsible for enforcing the safety rules. We all know what happens when you follow or break the rules. Discipline! Mangers will say something like, "We need to send a strong message to him (the injured) and all of our employees!" Translated, that simply means we want to use fear and intimidation to try to get everyone to comply with the safety rules. Issuing discipline is a trust killing action and it is understood in the Lean community that trust building is the key to Lean success. When I was introduced to the site manager, he informed me that the scheduled accident investigation meeting was not going to take place because the injured individual had not reported for work. I simply stated that I believe the focus of all accident investigations should be the process, rather than the person; therefore, we could proceed with the investigation. He agreed and the two of us headed to the Gemba and the work center where the accident had occurred.

We went to the grinding department where associates were using hoists to lift a variety of the welded parts onto tables where they would use disc grinders to remove welding flash and burrs from the parts. The site manager informed me that the individual had injured himself by placing his finger in a pinch point. As he was lifting a fabricated part, his finger was between the hoist hook and the part. We approached an associate performing the same task as the injured employee. As always, I began to ask a series of questions, a Lean tool known as "asking why five times," to get to the root cause of the process problem. I began by asking the site manager and the associate to be patient with me because some people can be offended when you pepper them with question after question. I asked the following questions. Why is a hoist used to move the fabrication? Why is the hook shaped as it is? Are all the fabrications the same? Why are the fabrications placed randomly on the floor rather than orienting them in a common direction or position? Where is the associate supposed to attach the hook to the fabrication? How is the associate supposed to hold the hoist hook? Where should his or her hand be located on the hoist hook? Why is there no handhold on the hook to keep the associate's hand out of harm's way? In a matter of minutes we had made the journey from "the injured was the problem because he put his finger into a pinch point" to "the process of designing the hoist hook had failed to provide a handle or grip point for the associate so that his hand and fingers could be kept out of the pinch points." The design process was the problem, not the person. I know it was an "ah-ha" moment for the site manager because I could see it on his face. If his organization practiced continuous improvement safety, rather than relying solely on a compliance safety program, someone would have probably recognized the hazard and changed the design of the hook prior to the accident.

I asked the site manager if he would like to observe another work operation to see what we could discover. We agreed to stay right where we were and watch the four associates who were grinding fabrications. He pointed out that there was a history of accidents due to contact between an associate's hands, arms, and legs, and the rotating grinding disc. I asked the following questions. Why are you using that type of grinder? Why is there no guard covering a portion of the rotating grinding disc? When does the injury occur during the grinding process? How long does the cutting disc keep rotating after the trigger mechanism is released? Where does the associate set the grinder when he or she has to move, change, or manipulate the fabrication on the table? Why does the table have a slick steel surface? Why is the grinder casing made of steel? What prevents the grinder from sliding, rotating, or falling off the table when the associate sets it on the table? The site manager, like most people promoted to these senior positions, was a very intelligent man. He now understood Lean Safety and understood that he had to provide the associates with a location that would secure the grinder when it was set down. I suggested scheduling a cross-functional team safety kaizen blitz event with the goal of eliminating injury risks in the grinding department as an outcome of our accident investigation. About this time, the HR manager, who was supposed to be part of the original accident investigation meeting, joined us in the grinding department. Apparently, he had gotten word that we had started without him. Together the site manager and I excitedly explained our "process problem" findings. We had not previously met nor did he know my background. HR managers are not generally process thinkers because they are often mired in compliance safety and routinely deal with people problems and therefore tend to focus on the person rather than the process. I continued to ask questions to educate them as we continued our walk. At one juncture, the HR manager stopped and said, "Who are you?" The way it was stated it sounded like "Are you from Mars?" I simply stated, "I am just a guy who sees safety differently and hopefully after this walk so will you." In reality, we were from two different worlds. He was from the compliance safety world and I was from the continuous improvement safety world. I think I shifted his orbit a bit.

Case Study 2

Electrical Cabinet Assembly Operation, Indiana

"This company struggled to see the connection between safety and their continuous improvement efforts. The Lean Safety workshop provided them with a very powerful tool to design safe, lasting improvements in their factory."

At times, I have to undergo 14-hour flights across the globe in order to share my passion for Lean Safety, and then occasionally I am provided opportunities right here in the Midwest. Living in the Chicago area allows me to drive to four other states in three hours or less. My home is about a 45-minute drive from the city center of Chicago. It is a wonderful city with a few drawbacks—the traffic and the winter weather—and many positives like its central U.S. location.

This Indiana site assembled components into electrical cabinets. I have been in at least four facilities on three different continents that do similar work. One of them is a very progressive Lean-focused organization. They have opened their doors to share their Lean journey story with others and I was fortunate to have attended a daylong workshop at their site in Chicago. What I observed there was in direct contrast to the site I was currently visiting.

In this book, I talk a lot about the differences between compliance safety and continuous improvement safety. As noted, compliance safety is a requirement driven by OSHA and a company's internally written safety rules and regulations. Because of my work history, I am aware of many OSHA requirements, but I am quick to tell clients and workshop attendees that I do not offer advice regarding compliance. I leave that to the educated safety experts who manage compliance safety programs. My goal is to help those safety experts see the other side of safety—the continuous improvement side. However, at this site compliance safety was lacking. Seeing this, I realized I could affect both compliance and continuous improvement safety during my time on site. Because this was a public workshop, there were a mix of safety and operational professionals in attendance. Rather than trying to keep the safety professionals reigned in and focused on continuous improvement safety, I asked them to also seek out and identify any noncompliance issues.

Standard safety practices, defined by safety compliance programs, help maintain compliance safety in an operation and standard work practices drive Lean improvement. What we observed on our Lean Safety Gemba Walk was a somewhat disorganized site with everyone trying to do his or her best without any standard work practices in place. Creating standard work practices makes the job most efficient and defines the safest way to perform a task. In my workshop presentation, I always make the distinction between standard work and work standards. Work standards are the "expected output from an operation." As an example, 20 parts per hour for a drilling operation is the work standard. The standard work for the same operation is "how you perform the work task." The task definition would define inventory storage locations, feeds and speeds of the drill, type of drill, part fixture, or clamping method, etc. The Lean Safety opportunities in an operation like this are positioning the inventory and equipment so that the individual working there keeps his or her body parts (back, shoulders, neck, etc.) in as neutral a position as possible. Armed with this information, we observed some of the assembly operations.

As we watched an assembler install parts into a large vertical sheet metal cabinet, we noticed he had to climb up and down a step ladder to perform the required tasks. While on the ladder, he had to twist his torso and bend his head and neck while trying to insert the required screws into the holes that coupled the part to the sheet metal structure. I began to ask questions. Why is the cabinet in a vertical position? Could it be laid on its side on an assembly table for final assembly? The components he was assembling ended up as a larger assembly in the cabinet, so the obvious question was "Why don't you assemble the smaller parts into the larger assembly on a workbench outside of the cabinet?" Many of the trips up and down the ladder were to retrieve the correct screws and parts required for assembly. When the Gemba walkers started discussing a new layout with inventory correctly positioned at assembly benches, I started backing away because they had begun to see safety differently.

I noticed that the workforce seemed a bit nervous as we engaged them in discussions. You could tell this was an environment where you did not ask questions or challenge how things were done. You just made do with what you had and did the best you could. The concept of stopping and watching a work process with the intent of improving the work for the assemblers was as foreign a concept as is the definition of the word "Hoosier" to anyone who lives outside of Indiana. In Lean terms, this was an unempowered workforce. By our being there, engaging the assemblers, the site managers standing beside me began to understand their responsibilities differently. The responsibility to take advantage of everyone's knowledge by engaging their reports in ongoing and never-ending discussions about how to improve the workplace operations is a business requirement today.

Why is that? One reason is that electrical cabinets can be assembled anywhere in the world, not just Indiana. The continuous improvement element of a business is as important as accounting, purchasing, quality, maintenance, or any of the other standard operational functions that define a business. By making continuous improvement a standard operating procedure (how you conduct and manage a business), any business can continually reduce its delivery cycle times to customers and reduce the cost of its operation. Speed to customer is the biggest advantage local producers have over their global competitors. If a business chooses not to improve, it is falling behind and in this very competitive world will disappear.

Once a business decides to pursue a continuous improvement agenda, it has to define the path forward. Using this site as an example, they were perfectly positioned to start by engaging their employees in discussions just like the ones that occurred during our Safety Gemba Walk. This focus on the culture, changing how people think, act, and interact, before you focus on the tools (Lean tools like 5-S, etc.) is the most direct route to Lean success. The easiest entry point to begin those conversations is Lean Safety—making work safer and easier while at the same time reducing the cycle times. A dialogue about keeping assemblers off ladders, standing upright with body parts in neutral, and inserting subassemblies with ease is how I would start the Lean journey if I were in charge of this business.

During the end of the day wrap-up presentation to management, the Gemba walkers presented well over 100 ideas that related to both compliance safety and Lean Safety. It was a very impactful one-day workshop that had the potential to influence the business over the long run. Many management teams resist admitting they have problems within their business to outsiders. They do not want to share their company's stories for fear of looking bad. When companies like this one have the courage to open their doors and expose their warts and flaws to others, wonderful results always occur. I shared this story, and all the others in this book, not to make anyone or any business look bad. It is so that others could learn from them.

Case Study 3

Distribution Center, Nevada

"Bob received very high customer feedback scores and achieved great results for the host company and the participants. Bob does a great job of keeping the class engaged and transferring new skills and knowledge to the participants."

"I really enjoyed the Lean Safety Gemba Walk, being on the floor and actually seeing and understanding the process really opened my eyes."

On this trip, I visited another area of the United States that was new to me. I found this part of Nevada to be quite different from the Nevada with which we are all most familiar—Las Vegas. The terrain was flat and dry and the population was sparse and very friendly. A short drive west took you up into the Sierra Nevada Mountains where you could access all of the outdoor activities the area had to offer. Some companies locate here, rather than in California, to gain access to the California market for their products. This was the case for the company I visited.

High-profile companies on your consulting client list give your work some level of credibility no matter what the results of your efforts. If you have watched the Daniel Pink YouTube video in which he discusses what motivates people today, you will better understand why people want to work for high-profile successful companies. In the video, he spells out his case for three motivations—autonomy (the desire to be self-directed), mastery (the desire to get better at something), and purpose. It is the purpose motive, the desire to work for a company that has an extrinsic value, to feel like you belong to or are part of something bigger than you are. This enhanced corporate image is a magnet used to attract not just good employees but great employees, and a company with that type of attractiveness can be selective in who it hires. Therefore, my host, the operations manager, and the other internal people with whom I dealt were professional and highly skilled. This was validated when all of the attendees at this workshop commented on the positive culture at this work site during our event wrap-up meeting.

I spent two days on site along with about 20 external visitors who had signed up for a two-day Lean Safety workshop. At the end of Day 1, we were given a quick walk through the facility—a huge distribution center (DC). Following the classroom portion of the event, I transitioned from my instructor role into my coaching role as we began our Safety Gemba Walk of the facility. I asked questions and interacted with the associates as we toured the DC. We could all see that a lot of physical work went on there as items were picked up, packed into cartons, and palletized for shipment to their customers. I was hoping the 20 visitors were as excited as I was at the prospect of having a positive impact on how the work was performed. I am always amazed at how all of us who work in facilities accept the work practices that exist there. We walk around with blinders that prevent us from seeing nothing more than people at work. We fail to give focus to the unsafe actions required to perform the work (a Lean Safety observation) that would help the associates and reduce work process cycle times. Instead, we take a monthly compliance-driven safety walk and look for either unsafe behaviors (compliance safety observation) that blame the associates or things that are noncompliant like blocked access to fire extinguishers. Compliance safety walks are a key ingredient in a compliance safety program and are a valuable tool in that effort. It is too bad most companies stop there and do not recognize the opportunities that result from continuous improvement safety activities like a Lean Safety Gemba Walk. This site leader was different. He opened his doors and asked for our help in making work safer and easier with the added bonus of reduced cycle times. He would not be disappointed.

On Day 2, the workshop attendees were split into four kaizen (continuous improvement) teams and sent to specific work areas to observe associates at work. They were given opportunity logs onto which they were to record as many opportunities as they could identify. I joined the site manager for a Lean Safety Gemba Walk. We observed individual workstations and checked in with each kaizen team to ensure they understood their objective and were harvesting opportunities.

The picking and packaging line had a linear flow with empty cartons placed on a roller conveyor to begin the process. The associates who picked the customer orders moved each carton down the line manually as they pulled inventory, based on the customer order, from the cartons and pallets located behind them. They inserted the material into the cartons, which eventually ended up at the end of the roller conveyor line. At this point, I observed an associate pushing each carton, and there were a lot of cartons, forward onto a powered roller conveyor that reversed the direction of the cartons and eventually delivered them to an overhead conveyance system that conveyed them to the associates who picked them up and placed them on the correct pallet for shipment. I started asking questions of the associate who was pushing each carton forward to the powered conveyor. Why do the cartons have to be pushed forward? Why do you have to push them? Would a mechanical pusher triggered by a sensor that detected the carton make your job easier? It was no surprise when the associate quickly responded with a "yes." To encourage him to think deeper, I asked him, "How could you add value if you didn't have to push the cartons?" His wheels were spinning as the site manager and I walked away.

I have had other managers, at other sites, respond to an issue like this by saying "That is why we use temporaries." They know the work is mindless work that does not value the individual and somehow they think it is okay for temporaries to perform the task. I believe all people want to add value and be valued by their employers. They want to make a difference and leave work each day feeling like they have made a difference. They also want to grow. They want to be more than they are today. If you are familiar with and understand the Toyota Production System (TPS), then you know there are three pillars upon which it is based. Delivery, the first pillar, is the requirement to push toward one-piece flow and all it entails, new plant layouts, material movement into and around your facility triggered by pull signals, like kanban cards, and a never-ending desire to reduce the cycle times of all of your business processes. The second pillar, jidoka, is the quality pillar. This pillar helps a business deliver quality parts and services and uses Lean tools like poka-joke (mistake proofing) to control the production or service delivery processes. The third and last pillar of TPS is respect for people. It is the one that most interests me and also the one most often forgotten by companies who begin a Lean journey. It interests me because Lean Safety is all about engaging the workforce in proactive safety improvement—it is my passion. Forgetting this pillar guarantees the failure of a business's Lean effort. It will fail because leadership neglects to engage, and therefore win the hearts and minds of their workforce, in the continuous improvement activity. When a manager anywhere tells me, "That's why we use temporaries" to perform work tasks not suited for his regular associates, I immediately understand that he is only partially committed to a safe workplace and does not understand the third pillar of TPS. At Toyota, they also believe that if you let an associate come to work day after day and perform the same task, you are being disrespectful to that associate. When I first read this statement, it took me a while to truly understand it. Eventually, I understood the inner meaning—by allowing someone to work at the same job day after day, you are denying him or her the opportunity to grow. So at Toyota people are continually challenged to grow. Today the statement, "We want to empower our employees" is heard frequently. What does that really mean? Hourly employees, when they initially hear the word "empowered" often think more work is going to be dumped on them; for example, they are going to be asked to take on the tasks that were once supervisions. If they have spent most of their work careers working in a top-down directive environment, their feelings are understandable. If they have only contributed to the business by using their hands to do the physical work, then discussions about empowerment can seem unclear and threatening. Yet a simple visual tool, when used to demonstrate the concept of empowerment, will move the conversation forward. Simply draw three squares starting with one and the other two progressively larger around the initial square. The inner square represents the tasks an individual performs when he or she is working. The individual does not have to consult any member of management or the support staff to complete these tasks. The decisions required for these tasks are made by the individual doing the work. The second, or center square, represents times when the individual may vary from standard work and he or she feels the need to inform someone, most often supervision, that a change was made. The outer and final square represents times when the worker seeks supervision's approval before making a change or an improvement to a work process. The goal, if you want to empower individuals, is to continually expand the inner square so that they can make more customer-focused business decisions. Explained this way, it is easy to understand that empowerment is really about decision-making or, more specifically, about who makes the decisions.

The ability to make decisions requires information. Therefore, empowerment requires the downward flow of information in an organization. Those who have traditionally been the least empowered in any organization, those performing the hands-on work, can easily make decisions on work scheduling and component replenishment if given the signals, or information, to do so. They can also suggest and implement cycle time improvement changes to their work processes when allowed to do so. By allowing decision making to be driven down in an organization, management broadens the scope of the work performed, which makes work more interesting and fulfilling. Employees are more likely to feel like they are "making a difference" because their new actions forge a stronger connection to the customer. Empowerment is a simple concept but it takes courage for management to let go of some of the decision making for it to happen. Now, back to the distribution center Lean Safety Gemba Walk.

This DC site leader was different; he valued his associates and was on the same page as I was because he quickly noted the opportunity to make work safer and easier for his associates by spending some capital dollars to automate the carton-pushing process. We then moved on where I observed a team of about 6 to 8 people removing cartons of product from a large metal shipping container. They put the cartons onto a portable roller conveyor, cut off the top of the corrugated carton, and then placed the carton onto a pallet. This was to make it easier for the order pickers on the packaging line we had just left. If you have tried to remove items from a carton with or without the lid, it can be problematic especially when you get to the bottom layer. I asked, "Why not carefully turn the box over and then set it on the pallet?" This would allow the order pickers to slip off the carton and expose product for easy access. However, the real question is why are they shipped in cartons with sealed lids? Could you use a carton with a fitted lid that would simply slip off so no one has to spend their workday cutting lids off boxes? Certainly, this opportunity fell into the "storage container" category of improvement opportunities.

I then moved on to observe one of the kaizen teams, which had been assigned the pallet-loading area. Here we observed an associate, whose job was to remove loaded cartons from the discharge conveyor and position them on one of four pallets alongside the discharge conveyor. Her first motion was to lean in and reach out to pull cartons from the far side of the discharge conveyor run out area. It was wide enough for at least two cartons, so to reach many of the cartons she had to move her back out of neutral. The kaizen team had estimated she would have to perform this motion approximately 200 times during her few hours of duty at the pallet-packaging station. On the opposite side of the conveyor was a deflector that, if adjusted, would drive the cartons toward her and minimize the extension of her arm and torso to get the carton. It was adjusted and the associate's facial expression reflected the understanding that work was now going to be easier. The important point is that for a long time everyone who worked at this station had to bend to some degree to pull the cartons toward them before they could pick them up. Simple small changes can make a real difference in how a person feels at the end of his or her workday. The team also recorded the associate's movement as she picked up cartons and walked to, and around, the four pallets to position the cartons on the correct pallet on a spaghetti diagram. The diagram showed she was spending much of her time walking rather than packing cartons onto pallets. This caused them to explore new layouts for the work area that would minimize her movements. Walking around carrying cartons is tiring and a safety risk because of the opportunity to trip and fall. A new layout that minimized walking distances would allow her to go home less tired each night.

That afternoon, the kaizen teams reviewed their list of Lean Safety opportunities and prepared their presentations to management during which they would talk about some of the more meaningful ones. As usual, I sat off to the side during the wrap-up meeting and watched the teams describe what they had observed and their recommended improvements. Combined, the four teams had identified around 100 opportunities to make work safer and easier. Often when a site is asked if they want to host a safety-related workshop, they shy away from the offer. This is because they assume it means a group of people will roam around their facility and point out a bunch of safety noncompliance problems that will make them look bad. After all, the Gemba is a reflection of those who manage it. Lean Safety events are completely different and our host site manager, who now viewed safety differently, graciously thanked all of the workshop participants for their help in creating a safer and more productive environment at his site. It was a win-win outcome. The site's safety culture had been impacted and the outside attendees left to go back to their own sites to make a safety difference.

Case Study 4

Transformer Manufacturer
Power Partners, Inc., Athens, Georgia

"This workshop had a central focus on safety, which brought a positive atmosphere to the shop floor."

The only time I had spent in the state of Georgia, prior to this visit, was in my car driving to Florida. Obviously, it is difficult to assess an area of our country and the people who live there from the inside of a car. This visit exposed me to a small college town, Athens, which was filled with people who were genuinely friendly and welcoming. College towns always have an interesting downtown culture that is a mix of kitsch and small-town businesses. I am a foodie so I often roam the streets of new towns like this looking for a particular type of restaurant. I love to cook gourmet meals for friends and family and have been doing so for decades. Therefore, when I travel to different parts of the United States, or abroad, one of the joys of travel for me is finding a local restaurant, with a creative cook, who gives focus to regional cuisine. On this trip, I was fortunate to find a small place where I enjoyed a great meal of regional southern cuisine specialties.

As a consultant, I have to find ways to connect with people to get them to open up. I always find that the topic of food helps me to easily develop a bond with the people I interact with during my consulting work. Food is a subject that quickly extends the conversation way beyond the weather forecast for the day. Food is the universal language and is one of the easiest entry points to begin a conversation with people from anywhere in the world. I have had some wonderful conversations with cab drivers about their favorite meals back in Africa, Pakistan, India, and elsewhere. They share their food stories with joy and fond remembrance evident by the huge smiles on their faces. Because I have probably cooked, or at least eaten, the meals they describe, we have an immediate and meaningful connection. As I pay the fare and depart, we say goodbye like old friends. Try it sometime and witness the joy it brings to people.

Often, when I visit a site for the first time I am little nervous. My concern is the availability or quantity of Lean Safety opportunities present in the facility. I want to ensure that my client and the workshop attendees truly understand Lean Safety. For this to happen, it requires a site where people are physically working. I have been in some plants with very little physical activity taking place and yet there are always good Lean Safety examples. Upon my arrival at this site, I was immediately ushered into a meeting room where I was introduced to about 15 managers from all parts of the business. I talked about my background, attempted to explain why I had been invited in, and answered any questions they had. I then began a Lean Safety Gemba Walk with four of the staff. As soon as I walked onto the Gemba, any concern I had about the availability of good Lean Safety improvement examples vanished immediately. I witnessed some very physical and demanding work. This site made components for and assembled electrical transformers. Even though this was my first time in this plant, I felt as if I had a connection to the culture.

The reason for that is I had known one of the business owners through the Association for Manufacturing Excellence (AME). She, like me, was an AME volunteer and a person who genuinely cared about people. When she and a partner purchased this business, it was about to go under and if it had, all of the jobs it provided would immediately vanish from the community. Together they rebuilt the business using the respect for people pillar of the Toyota Production System (TPS). I knew this because of some testimonials I received from people who worked there. That occurred at an annual AME international conference—a gathering of approximately 2,000 Lean thinkers that takes place each fall. One of my volunteer roles at the conference was to facilitate sessions in the Idea Exchange Café. These were sessions open to all with only an advertised topic, such as "How Do You Engage All Employees in Your Lean Efforts?" to attract attendees. As people walked into the room, they took a seat in a circular layout. I would then guide, or facilitate, their conversation for the next 60 minutes. Power Partners had begun to send a large contingent of associates to this conference. Many companies only send managers and technical staff to costly conferences. This co-owner of

Power Partners was different. She understood that if you want your employees to think like you do, you should expose them to the same experiences. Therefore, her team had associates from all areas and levels of the business. As it turned out, she and some of her production staff were in one of the Idea Exchange Café sessions I facilitated. Although this happened about four years ago, I can still vividly remember what transpired. A production associate from Power Partners began to share how the culture of the business had changed for the better and the person responsible for that change was the woman who purchased the business. He talked from the heart about how respectful and trusting she was and how being treated that way had personally affected him. As he talked, I glanced over at the owner and saw her tearing up. I think others in the room might have been doing the same. She, like me, understood that the greatest reward for leading Lean is the impact you have on people. She was a people-centric leader. Therefore, when I visited this plant I knew a bit about the culture and could see reminders of the owner everywhere even though she had passed away a few months before.

I spent a total of three days on site. After the initial Safety Gemba Walk, I observed a safety team meeting followed by an incident investigation. The incident that was investigated involved a forklift that had slid on a wet floor in the shipping area of the plant. The initial focus of the investigators was the forklift driver and the speed at which he was driving. It was implied that he should have slowed down because he knew the floor was wet. By asking why five times I tried to steer them away from the person and toward the real root cause. Why did the forklift slide? Why was the floor wet? Where did the water come from? How can you contain the water? Why do the forklifts have to go outside? What do the forklift tires look like? Are there different tires that would be more suited to wet conditions? The task list that resulted from this investigation had items related to tire selection and water containment on a ramp used by the forklifts when they returned from outside. I helped them focus on the "what and why" and not the "who."

Someone other than the forklift driver most often reports forklift incidents. When I ask workshop attendees why, they quickly point out that it is because the drivers fear getting in trouble. I recall in the manufacturing plant where I worked, because of a large number of forklift incidents, the training manager was investigating installing impact monitors to our forklifts. They would cause the forklift to shut off if it impacted anything with a force above the set limit. The forklift driver would then have to contact the supervisor who possessed the key used to reset the forklift impact monitoring system. My reaction to this proposal was not very positive because I believe load monitors are trust killers. In our plant, as in most, the individual who found the damage, rather than the driver, had reported the forklift incidents. This was an obvious signal that they did not trust management or at least our response to the incident. I suggested that rather than police our forklift drivers with load monitors we investigate every incident that is reported to help us understand why the incidents are occurring. I engaged a forklift driver in every incident investigation. Together we discovered aisles too narrow to swing a forklift and its load into the desired location, equipment in the way of the swing arc required to load and unload machines, and signs and items hung off walls and ceilings that interfered with the forklift movement. Because of our investigations, we began to change our facility to make it easier for the drivers to do their job. In return, they extended trust because they started reporting incidents when they occurred. They did this because they knew they would get help, not discipline. Management has to extend trust to get it back. I shared this story with the incident investigation team hoping in the future they would continue to focus on the "what and why," and not the "who" during their accident investigations. If they did, they would be helping to grow the level of trust in their plant.

On Day 2, approximately 24 associates attended the first day of my Lean Safety workshop. We covered my slides and they participated in multiple small team exercises. As usual on Day 2 of the workshop, the attendees were split into teams and sent to the Gemba to find improvements that would make work safer and easier. One of the four teams went to an area where formed sheets of steel were assembled into electric coils. Watching the associates complete this work made my back hurt. They were hunched over a large heavy object and with their hands used a playing card shuffling movement to alternately layer the formed steel plates on the coil assembly. I was very impressed with their strength, ability, and agility. The team did a great job of investigating new methods, tables, table heights, hoists, etc. to reduce the injury risk associated with handling such a heavy part. Sometimes when I depart a facility, I am not personally satisfied with the results of our efforts. This particular work task was one of them. We did not have enough time to make a real difference. I still think about this work task and hope the internal staff continued to seek real meaningful solutions and if my travels take me close enough, I will stop in to see. Easy solutions are not always apparent or possible. That is why it is called continuous improvement.

The four safety kaizen teams contained employees from this site and another company-owned site located nearby. Each had a higher percentage of shop floor associates than I usually see in my workshops. Their uneasiness helped me understand they were not accustomed to participating in workshops or standing in front of others giving report outs. Credit goes to Power Partners for including them. I can tell you it made my time there more rewarding because I was able to help grow and build their confidence in their ability to make a safety difference.

When I contacted Power Partners to ask if I could include the company name in the book, they agreed and also informed me that changes had occurred in the coil assembly area as described in this case study. "We've converted one entire line that builds core/coil assemblies into three cells. Each unit that is built now rides on its own cart, instead of down a conveyor where there was a lot of pushing and pulling of units. Each cart can be raised and lowered to the needs of the operator he or she is working on the unit. It doesn't eliminate the heavy product work, but at least the operator can adjust the work piece to allow a more body neutral stance than before and be closer to it." It was rewarding to hear that they did indeed continue to give focus to this physically demanding work area. This is a great company because they understood from the very beginning of their Lean journey the importance of focusing on culture first. "Culture first and the Lean tools second" is my mantra. Too many companies focus only on the Lean tools and fail to ever impact their business culture. Those are the companies that are implementing 5S for the third, fourth, or fifth time.

Case Study 5

Industrial Distributor, Pennsylvania

"I became a believer in Bob's concepts after reading his book on Lean Safety. I was further convinced that he could impact safety and even productivity after attending an AME workshop that he facilitated. I then hired Bob to do an event at my company and he impressed me and many others with his ability to quickly understand our environment and provide and foster out-of-the-box thinking about how the job was done. I highly recommend Bob and his concepts. His work is truly making a difference."

This visit took me to an industrial distributor with small sites located across the United States and Canada. I was to begin my visit at their headquarters in Pennsylvania. I had not visited Pennsylvania since the late 1970s but feel like I have a connection to that region of the United States. That is because it was the epicenter of steel production in the United States for decades, and the company where I was employed for 18 years, U.S. Steel Corporation, was headquartered there. I did not expect to see much of the local area because I was scheduled to fly in one evening and then depart the following night. As it turned out, I arrived late and departed 24 hours later than expected. Both delays were the result of airline issues. So, I did see and experience more than I wanted of the airport and an airport hotel but very little of the client site city.

My initial interaction at the headquarters was to provide a Lean Safety overview to about six people including the executive who had hired me and his supervisor, the COO of the business. My initial conversation with my company contact had been about four months earlier. He had a strategic role in the business and one of his objectives was to turn around (improve) their safety program results. He called after he had read my book and wanted to know more about Lean Safety and its applicability to an industrial distribution business. He explained that because they sent service crews into large customer facilities, those firms often evaluated their safety record. Of late, their accident rate had been quite high and there was some concern about losing a large service contract. One of my suggestions for him was to attend one of my upcoming public 2-day Lean Safety workshops. By doing that, he would be exposed to the Lean Safety philosophy on Day 1 and the hands-on application of those principles on the Gemba during Day 2. Some weeks later, I had the opportunity to sit and talk with him over dinner after the first day of a workshop held in Nevada. He told me more about his business, small service/inventory sites with staffing levels from 12 to 40 people. He was struggling with how to create a culture of safety standards within this dispersed business model. Each of these sites had its own business culture and he clearly understood that someone from corporate headquarters walking in and giving orders would obviously not work. For instance, multiple injuries related to using box cutter knives had occurred over the past year. He wondered if he should select one knife and tell everyone that was the only knife they could use. He shared with me their injury history report, which was a flashback to the past for me. One of the corrective actions listed for many of the accidents was to "retrain the employee." When I was a young supervisor at U.S. Steel, I used this reason code myself. This statement tells you nothing about the real root cause of the accident. It is akin to another reason that often preceded the retraining statement—"the employee is sorry and said it won't happen again." That statement implies that the employee is the problem and if we just retrain him or her, the problem won't happen again. It is a lazy attempt at accident investigation. It is a statement you can write while sitting in your office without ever talking to the injured or going to the Gemba to seriously investigate the accident. Today I understand and I wanted to help his entire multi-site organization understand that by asking why five times you could get to the real root cause of an accident. Accident investigation meetings are continuous improvement meetings. They are an opportunity to get to root causes and real meaningful corrective actions. By focusing on the "what" and the "why" and not the "who" during the investigation meeting, you begin to build trust. You can then build upon this trust to start changing your culture from one that blames the injured for the injury to one where you focus on safety process improvement. Using the statement, "retrain the employee" implies that the business has standard work processes in place that everyone follows and the injured individual had deviated from the standard. That was not the case in this business as I found out soon after I completed my training session in the headquarters building. Next, we traveled a short distance to their local production operation to take a Lean Safety Gemba Walk.

I simply guided the managers on a walk through their facility and as I did, I engaged the associates doing the work in discussions about improving how they performed the work. The Gemba, along with the attitudes of the associates who work there, was a reflection of the managers who manage the site. I found the associates open to discussing change, but I also noted an implied understanding that management calls the shots around here and they just do the work. This is not unusual and it is simply a reflection of an unempowered work force. I see this as an untapped creative force merely waiting to be asked for their ideas. Once I engaged them in discussions, the managers that I was guiding quickly engaged in the continuous improvement conversation. These clearly were not the normal everyday conversations that occurred at this site.

One work process I briefly described in an earlier chapter had to do with installing fittings on the ends of a very large piece of plastic tubing. When we arrived in the work area, the 20-foot-long piece of tubing had been uncoiled on the concrete floor. The first step, as described earlier, was to remove a plastic overwrap from the entire length of the tubing. This was a prime example of a supplier providing more quality than the customer required. It is one of the seven deadly wastes—over-processing. Often, marketing and engineering professionals will decide what their customers need without input from their customers. On the other hand, they will add a standard feature to a product offering because a few customers have requested it. This probably fell into that last category because covering the entire length of tubing with a plastic cover that had to be cut off by someone crab crawling along the floor with a box cutter knife was not only a waste but also an injury risk. The tubing had been shipped in a corrugated carton and needed no further protection. My guess is that in the past a customer called and complained about the tubing being dirty when it was received. Sales, marketing, and engineering collaborated to solve the customer problem by adding the covering to all tubing. I suggested they contact their supplier who could then contact the manufacturer and ask for a change. Good intention changes to products do not always translate to "value added" changes. The next steps in processing this large piece of tubing was to insert large metal fittings into the ends and then position the tubing ends into a hydraulic crimper to crimp the fittings. Inserting the fittings was a wrestling match and the hydraulic crimper was so high the associates had to reach up to shoulder height in order to manually hold the heavy unwieldy tubing in position while it was crimped. After my questioning, all agreed scheduling a multi-day kaizen event to reduce the injury risks associated with this task was a great idea. I would expect the results from this event would be a workstation that would provide clamping stations for fitting insertion and the lowering of the crimper to bench height. We then moved to another area.

As we approached an area where they processed small hydraulic hoses, we observed an individual cutting hose to length using a chop saw. He was using his hands as clamps because they held the hose in place during the cut and the cutting process was initiated by his hip, which pushed on a plate that moved the blade through the hose. People that worked in this facility only observed someone at work. I observed hands at risk, a back bent out of neutral, and a body part used to drive a machine process. Often what happens is that crude rudimentary manufacturing process steps are quickly defined to process a few pieces of some new or low volume product. Businesses are reluctant to spend capital dollars on the best, maybe automated, process when there is no volume to justify the expenditure. Then the volume increases and no one goes back to redefine the manufacturing process. My guess is that this process fell into that category.

We continued on our Safety Gemba Walk and viewed two additional work areas—one where they were crimping fittings that had been pushed into small-diameter hoses and a second where large rolls of material were being slit. In both areas, additional opportunities were identified. This Safety Gemba Walk was intended to open the eyes of all who participated to the almost endless opportunities that existed in their facility to make work safer and easier and drive cycle time out of their customer service processes at the same time. They also learned that if they want a long-term reduction in injuries, they would have to engage their employees in the continuous improvement of safety. Based on the quote from my company contact at the start of this case study, my mission was accomplished. Let's move on to another site.

Case Study 6

Mining and Metals
Metinvest, Donetsk, Ukraine

> "Robert did a great job delivering a Lean Safety seminar for top management of my company. The concepts and ideas were conveyed to the audience in a concise yet comprehensive manner leading the participants to view familiar issues in a new light and inspiring them to use new approaches."

I talked about some of my experiences in the Ukraine in an earlier chapter about BBS. The following includes some of the same material but gives a more complete overview of my visit.

I was invited to the Ukraine by a company called Metinvest, a large company with approximately 130,000 employees working mostly in coalmines and Soviet-era constructed steel mills. Metinvest's safety professionals and corporate communication staff planned and skillfully executed a safety conference titled, "Good Safety = Good Business." It was a two-day event attended by approximately 250 people from 70 different companies located in both Eastern and Western Europe. Noteworthy was the fact that Metinvest not only planned but also covered the cost of the conference—attendees paid no conference fee. The venue for the conference was a new modern hotel, the Shakhtar Plaza, in Donetsk, Ukraine.

Donetsk is a coal-mining town founded by a Welsh businessman in 1869 who extracted wealth from the region via steel making and coal mining. Donetsk is located in southeastern Ukraine near the Russian border. Russian, rather than Ukrainian, is the language spoken here. It is slowly changing from a drab Soviet-era town to a town with features to be proud of like the newest and largest soccer stadium in Eastern Europe, Donbass Arena, home of the Shakhtar FC. Springing up on the outskirts of Donetsk are modern retail and grocery stores and, I am sorry to say, the almost universal sign of progress for small towns and emerging economies, two McDonald's restaurants. This is a city and a country moving forward while defining a new identity in the post-Soviet era.

My Metinvest contact, with whom I worked for months to plan and arrange my visit, was a Russian-born safety professional who is passionate about making a safety difference in a country and industries where safety was not a priority. For example, the fatality rate (fatalities per 100,000 workers) in the Ukraine is around twice that of the United States. She and her safety co-workers were former "DuPonters"—individuals trained in the DuPont behavior-based safety philosophy. It was explained to me that the DuPont methodology requires you to work with your people to get to zero injuries because it is mostly people who make mistakes leading to injuries and it is people who remove barricading, fail to use lockout procedures, take shortcuts, etc. This is a sound approach to safety management and yet I feel this methodology gives focus to safety from only one perspective—the worker's activities relative to safety compliance. At first, the DuPont method seemed at conflict with my "the process is the problem, not the person" Lean thinking approach because I believe in eliminating the hazard when possible and thus eliminating the training and auditing required to help the person avoid the hazard. Of course, there is not one right approach, only different approaches that all have merit. For instance, both my method and the DuPont method focus on employee engagement but for different reasons. My mission, however, was to help the conference attendees understand the Lean continuous improvement employee engagement approach to proactive safety improvement. What is different and, dare I say, new about my approach is that I ask people to only watch people work and make their job safer and easier. It is not a focus on adherence to compliance requirements; it is a focus on the continuous improvement of the individual's work activities.

Prior to the start of the conference, I was given a tour of two parts of the Ilyich (Lenin) Iron and Steel works. This over 100-year-old facility has quite a history. For instance, two of their blast furnaces were disassembled and moved to northern Siberia during WWII to prevent them from being destroyed by the invading German army. Metinvest purchased this facility just two years ago and has only started to make changes. Today, this drab aging site employs around 35,000 people. I was taken on a tour of the steel making and slab casting facilities before making a second stop to tour a plate mill that produced steel plates used for shipbuilding. Since 18 years of my work history were spent at U.S. Steel Corporation, I was familiar with the processes I viewed. No evidence of Lean or continuous improvement activities was visible during my two stops. Leadership seemed stuck in the top-down directive Soviet past. This in no way reflects badly on them for they are a reflection of their culture and working environment. Change is indeed the only constant and we all have to face a changing world both at work and in our daily lives.

On the afternoon of the second day of the conference, I joined a bus full of conference attendees to tour the Metinvest Khartsizsk Pipe Plant—a facility purchased by Metinvest in 2006. This site manufactures large-diameter pipe for the oil and gas industry. The contrast between this plant and the Ilyich plant was stunning. I witnessed a completely different work culture and work environment due to application of the 5S Lean philosophy. This was the largest facility I had ever visited that had implemented 5S. The young, proud, and Lean-savvy leaders at this plant talked about how it was when they first arrived. Windows at the roof level were missing and workers would stand around empty 55-gallon drums of burning wood to keep warm. They transformed a facility that was filthy and disorganized into a bright, clean, and organized plant they were proud to show us. It was a great example of what determined focused leaders could accomplish. I was very impressed.

Prior to my trip I had read that Ukrainians are stoic and smiles are rationed like consumer goods during their Soviet past. This was true if you were passing someone you did not know, but when I was engaged in a conversation or when they talked among themselves they were just as friendly, animated, and prone to smiling and laughing as friends back home.

I found them to be warm, friendly, and genuine people. They expressed some concerns about how they and their country are perceived by westerners. My opinions are distilled from my experiences and all of my experiences in the Ukraine were very good. Building a new society is a big job, but the Ukraine and the determined people I met are on that arduous journey.

Not speaking their languages (Russian and Ukrainian) was certainly an issue for me. Not only is the language different, but also the letters of the alphabet are Cyrillic. As I walked around Donetsk, I had no idea what most retail outlets sold because I could not read the signage. Fortunately for me, the conference organizers provided simultaneous translation into either Russian or English. The translators, who did a wonderful job, were present in all of the conference sessions and on the plant tour. My role at the conference was to present a 40-minute keynote and a 3-hour workshop. Lean concepts were new to most of the conference attendees, so my keynote began with an overview of Lean before making the connection to Lean Safety. I challenged them to focus on process, people engagement, and the continuous improvement of safety. Since most of the attendees have worked or still work in cultures that rely on top-down direction, the concept of employee engagement was as foreign to many of them as the Russian language was to me. The workshop was for a select group of Metinvest site directors and a few other senior leaders who were attending the conference. Many of their plants employ thousands of people, so influencing them could lead to meaningful safety change. With that end in mind, the workshop included small group exercises that engaged them in defining both an employee safety engagement survey and leader safety standard work for themselves. Standard work is a Lean tool that has gained wide acceptance as a method to get to the culture and drive meaningful change. Standard work is a repetitive task that if performed routinely can change the behavior of an individual, which will in turn be recognized by those in contact with that person. An example is the requirement for a production supervisor to visit the team board in his area of responsibility, talk with his production team, and record the output against standard for the prior two hours. If they are behind goal, he can practice a coaching style of leadership by asking the right questions to determine why the team is behind schedule and what their plans are to get back on track. He should also offer his assistance if they require it. Again, his goal is to ask the right questions, not have the right answers. If this is done routinely, day after day, week after week, it will indeed be standard work not only for the supervisor but also for the team because they have to respond to his visit every two hours. I asked the Ukrainian managers to make a list of leader safety standard work. I challenged them to think of routine actions they could take that would send a strong message about the importance of safety to their workforce.

Because I was presenting in English, the workshop attendees, who mostly spoke Russian, wore headsets and were intently listening to the translator located in a booth at the rear of the room. For me, the lag time between what I said and when they heard it from the translator was a bit of a disconnect with my audience. However, it was not as bad as presenting a webinar, which requires you to talk to your computer and never see your audience. As they stared off into space listening to the translation, I attempted to make eye contact and read their delayed body language to help me understand if my message was getting through to them. When planning my visit, my contact had made it very clear that the predominant management style in use was a top-down directive approach. She knew my message about employee engagement would challenge their thinking and it is one of the reasons she invited me to present.

Safety professionals also use a top-down directive approach. They "push" safety down through the organization and are often seen as safety police who enforce the safety rules. "Push" is another Lean term that refers to how you manage your production process. If a company's sales forecast history is used to drive the products and the quantities of those products through their production facility, then they are using a "push" production process. A company that functions this way relies on enterprise resource planning (ERP) systems and work orders to manage that business process. The Lean community views ERP systems and this push production process as a source of "waste" because it results in not only excess inventory but also often the wrong inventory. One of the goals of Lean is to introduce "pull" production processes into a business. The pull production model simply replenishes what the customer has purchased based on some type of signal to the upstream production process. If you have ever visited a Subway sandwich shop, you will quickly understand what I mean. When you walk in the door, you are a pull production signal. You walk along the production process and pull from inventory the components to build your sandwich. Years ago, when McDonald's only had a few product offerings on its menu it used a push system. Prior to the lunch rush, employees would build inventory of hamburgers and cheeseburgers, based on their sales history, and store them in a heated cabinet. This allowed them to quickly serve the customer. However, as with all push system inventory processes, they would get the forecast wrong and product would sit in the steam cabinet too long. If you came in at the end of the lunch period, you might get a cheeseburger that was stuck together and dried out. Therefore, to ensure their customers received a quality sandwich, McDonald's switched to the make-to-order, pull system model.

Lean thinkers do the same when the Lean journey is started in any organization. They have to push Lean onto the organization and the individuals who work there. Eventually, they build a cadre of Lean believers and the effort takes on some momentum. Their success is dependent on the organization starting to pull Lean. What do I mean? When individuals start to ask the Lean staff for support to implement Lean tools in their area of responsibility, they are pulling Lean. This signifies that Lean is starting to become cultural. Now, think about safety professionals. They are part of a hierarchical top-down directive push process. It originates with OSHA, or some other regulatory agency, and cascades to organizations, safety professionals, managers, and finally the people who do the work—the individuals in harm's way. This is why they are often viewed as safety police. To change that image, safety professionals and managers have to engage their workforce in safety improvement activities. If they do, they stand a chance of creating pull in their safety program. Those in harm's way may ask for a safety kaizen event in their work area to make their work safer and easier. If this occurs, they will view their safety manager and supervision not as safety police but as safety resources. This is one of the messages I tried to convey to my Russian-speaking audience.

Overall, it was a wonderful opportunity to make a safety difference in a country where historically safety was not a priority. Metinvest, by hosting their second safety conference, is helping to lead the Ukrainian nation on a journey to make workplace safety a priority. I hope to be invited back to a future conference to witness their progress. More than one person said, "See you next year," and I enthusiastically replied, "Da (yes)!" The most rewarding part of my new career has been the opportunity to meet and get to know great people all over the world. My primary contact, a true safety professional, and all of her Metinvest associates are making a real safety difference not only at their company, but also in their country. I was honored to be invited and have the opportunity to contribute to their effort to make Metinvest and all workplaces in the Ukraine safer places to work.

Case Study 7

Food Plant, Australia

"Robert is a true facilitator of learning. He took the workshop participants on a journey of change over two days that led to new insights and understanding of the safety risks in their workplaces and how to use Lean tools to eliminate them. Participants left with knowledge and some experience of participating in a safety kaizen process that will allow them to take what they have learned and apply it in their own businesses. The result will be safer and more productive work places and more engaged employees. I would highly recommend Robert Hafey to any business or consortium looking for a way to reinvigorate their Lean efforts, to engage their people, and to eliminate the hidden safety risks in their workplace."

One of the sites I visited during my second trip to Australia was a food plant. They hosted a two-day Lean Safety workshop. The SIRF Roundtable organization was the sponsor of the event and the host site was a member of a consortium organized and managed by the local SIRF facilitator. As with all two-day workshops, at the end of Day 1, we were given a quick familiarization tour of the entire facility. Food plants, in addition to all of the standard workplace safety regulations, have to comply with food safety regulations. This facility was impeccably clean and organized and before we could enter the production areas, we went through an elaborate process to ensure we understood our responsibility to maintain their high standards. The last step before entry into the production area was to don protective clothing that covered each of us from head to feet.

Large food plants today are highly automated process production facilities, which helps to ensure product uniformity and quality along with minimizing the amount of labor required. This type of plant causes me some anxiety concerning the quantity of opportunities that might be present for the Lean Safety Gemba Walkers to observe. Will there be enough good examples so they can learn and then take it back to their facilities? If all we see are a few individuals watching control panels and gauges, it could result in an abbreviated Safety Gemba Walk and a blank opportunity log! What I have found in automated plants is that there are some specific points at which the associates have to interact with the automated processes and that is where the Lean Safety opportunities can be found. The Safety Gemba Walk should focus on tasks like loading packaging materials or products, removing completed products, change-overs, or troubleshooting the equipment.

One of the first areas we visited was the beginning of a fill and packaging line. Located there was a pallet of flat corrugated cartons. The pallet load was covered with spin-wrap plastic that was applied by the supplier to protect the corrugated cartons and to keep them from shifting on the pallet while in transit. When the associate was ready to remove the plastic wrap so he could access the cartons, he pulled what could be best described as a butcher knife from a sheath and cut the wrap. I was thinking, "I hope his supervisor never gets him really angry!" Spin wrapping pallets is one of my hot buttons and I will tell you why. I find there are no standards for how many rotational wraps should be applied. Companies, once they buy a spin-wrapper, tend to wrap every carton. The associates using the spin wrapper think the more wraps the better. I have seen pallets so well wrapped it appears the product is stuck in a spider's web. Trying to cut it all off is tedious and fraught with danger. Cutting off plastic spin wrap material requires a knife—think safety hazard. The knife can also cause damage to the product. The plastic wrap, the labor to apply it, and the labor to cut it off add cost but not necessary value. Finally, the plastic wrap is put in the garbage and a business has to pay someone to haul it away and put it in a landfill. Therefore, rather than just assume you must spin wrap every pallet, ask your customers or suppliers what the alternatives are. I also have a product idea to reduce both the safety risks and the cycle time to remove the wrap if you have to continue using it. I would call it the "Pallet Zipper." It could be an extruded roll of plastic with the profile of a block C with thick walls. You would simply cut a piece to match the height of the product on the pallet and then staple one end to the pallet and tape the top end to the product with the C facing outwards. After spin wrapping, this brightly colored plastic "zipper" device would be very visible. On the receiving dock, you would simply slip the blade of your box cutter knife into the opening of the C shape and, just like opening a zipper, cut the plastic wrap off in one simple motion with no danger to yourself or the product on the pallet. If you turn this product idea into a viable product, you owe me a commission of some sort!

As we continued our walk along the pack and fill line, we engaged an operator by asking if there was anything that could be done to make his job easier and safer. He quickly pointed out that he had to repeatedly lift very large and heavy safety interlocked doors to access the working areas of the packaging machine. The team suggested installing air shocks, like those on the tailgates of most SUV vehicles, to assist in the lifting process. The item was added to our opportunity log and we moved on. As we were walking, I thought about a common and very serious safety violation that occurs whenever you have interlocked safety doors intended to protect the associates from the moving and dangerous processes inside of a machine.

A common technique used by machine operators is to put a piece of tape or a magnet over the door safety interlock switch so they can open the door without shutting off the machine. This category of safety violation often results in discipline for the operator if it is discovered. I often engage my workshop attendees in a discussion of this type of violation by asking them why operators do this. I preface the question by stating that nobody gets up in the morning and informs their spouse over breakfast, "You know, today I am going to put tape over the safety interlock switches on my machine." The most common response to my query is, "Because they are in a hurry." I disagree with them and state that my belief is that they do it because they are frustrated. Frustrated with the machine or the tooling because it does not function the way it is engineered to and they have to repeatedly open the door to access the machine and tooling, and every time they do it stops the machine. Out of frustration, they tape the safety interlock. This is another example of the process is the problem, not the person. Fixing the mechanical problems that exist within the machine will remove the frustration and the need to open the interlocked doors. I was hoping they would install the air shocks on the machine we just observed before the operator tapes the interlocks and leaves the doors open because he is tired of lifting them.

The next day, the workshop attendees were split into kaizen teams and sent to specific areas to engage the workers and identify opportunities. One of the teams worked in an area where there were large stainless steel cooking vessels. They had to bring the different ingredients up to the elevated platform where they worked and dump them into the cooking vessels. With the operators, they identified multiple improvement ideas regarding the material handling of the ingredients. While I was visiting the team, one of the supervisors noted they were adding a cooking vessel to the platform, which was already a rather restricted and tight workspace. With a lot of pride, he then informed me what one of the operators had done. Using corrugated sheeting and tape, he had created a life-sized model of the top of the cooking vessel just as it would appear on the platform. This allowed them all to move the model around to determine the best location from a material handling perspective. Brilliant! I asked to be introduced to the operator who guided me to its storage location so I could see his creation. As he stood there beaming over his creation, it reinforced my belief that everyone has ideas, we just have to provide people an avenue to express them.

Case Study 8

Chemical Plant
Nufarm Limited, Victoria, Australia

"The Lean Safety experience was a refreshing extension on safety thinking and safety programmes. Our staff were engaged with the process and found both the subject matter and presentation style of Bob Hafey a rewarding exercise. Interaction with other safety professionals and plant staff in this programme provided stimulus to challenge our safety paradigms."

"I attended a two-day Lean Safety workshop facilitated by Bob in Melbourne, Australia earlier this year, and found it a very worthwhile experience. Over the course of two days, Bob showed that safety improvements can be a by-product of kaizen activities to improve quality, efficiency, or productivity; by applying Lean thinking to a safety kaizen event, an organisation can still achieve the same positive outcomes with considerations made to the team members instead of just the process. This type of thinking builds the foundation for an engaged work force and a positive safety culture. Bob's passion towards safety improvement is definitely contagious as he engages and shares his wealth of experience in applying Lean thinking across various industries. I am keen to incorporate the learnings from this workshop into my activities with the organisations that I work with!"

Once again, this was a workshop open to consortium members and held at a consortium member site—a chemical process plant located near Melbourne,

Australia. This was the first time I had a site manufacturing manager in attendance for the full two days of the event. He was very keen to redirect his plant's safety program and took advantage of this opportunity for some new learning. I am not sure he had always envisioned being on one of the kaizen teams on Day 2, but our Safety Gemba Walk, at the end of Day 1, solidified his decision. As we walked and I engaged his workforce with questions about their everyday work, I could see his interest peaking. Four kaizen teams were formed on Day 2 and he was a full-time member on one of them. The four teams were assigned one of the following work areas to evaluate and improve:

1. An automatic packaging machine that had recently been brought on line
2. A manual labeling station for small plastic containers
3. A flatbed truck platform where returned containers were unloaded
4. A plastic tote tank wash station where returned containers were cleaned out

The site manufacturing manager was on the team that was assigned the wash station. Much of the product produced at this site was put into large plastic totes that were stored in tubular steel frames to protect the container during movement and use. As the containers and frames were returned from the end users, both had to be washed prior to being refilled. A square piece of steel was on the tubular frame onto which the adhesive-backed product identification label was placed. On our Safety Gemba Walk the prior afternoon, we observed an operator hunched over a metal frame, using a hand scraper, attempting to remove the old label. We have all been frustrated trying to remove labels and the remaining adhesive residue from new products we purchase and this associate lived with that same frustration all day long. When I observed him I asked, "Why is the frame on the floor, which requires you to bend over to remove the label? Can you position the frame on a stand so you stay upright with your back in neutral? Why are you using that type of a scraper? Would a different tool, maybe a pneumatic scraper, make your job easier? What adhesive is on the label back? Would a different adhesive make it easier to remove the labels? Why do you have to use an adhesive-backed label? Can the product be identified with a different method?" All of these questions were to challenge the status quo and stimulate thinking. The kaizen team approached these opportunities along with many more in the wash station area and in the end made marked changes that made work safer, easier, and more efficient. The results were a win–win for both the associates working in the area and the company. That is what impressed the manufacturing manager.

The team that was assigned the plastic bottle hand-labeling area engaged a long-term employee in their quest to make changes that would make her job safer and easier. They ran into resistance. The associate did not seem interested in their suggestions that centered on developing a one-piece linear flow line. Her current process was to work on batches of containers by applying all of one label and then handle the same batch to apply a second label before moving the batch to the pallet. Triple handling of the containers was deemed both waste and a safety risk by the team. Despite her lack of support for their ideas, they moved the tables and other equipment required into a one-piece flow line that would have the operator complete one container at a time before she positioned it on the pallet. The changes would obviously keep her more neutral and reduce the amount of material handling required to label each container. They provided training on how to work in the new arrangement and even performed the work tasks to understand them and make a connection with her. One of them even simulated being an automatic label dispenser by handing her the labels so she did not have to peel them from the backing. I had been in and out of this work area as I bounced between teams to check on their progress. When I watched her, I could see her body language was as clear as a harvest moon on a clear October night. They were messing with her stuff and she did not like it. I have been in this position many times before and what I have learned is that the importance of the improvement to the individual who performs the work task determines its probability of long-term acceptance. This is especially true in a work environment where the workers have not been trained or exposed to Lean concepts and principles. That was the case here. If the team had made a dramatic change that would have eliminated a strenuous task from the work process, she would immediately have accepted the change. Because the team's recommended changes had to do with changing the sequence of what she did, she could not grasp the value of the changes for herself. She liked doing it her way because that was the way she had always performed this work task. The best you can do in these situations is to provide some training on the Lean concepts and explain multiple times why the changes are ergonomically good for the operator. Then you have to look the person right in the eyes and ask if she will work with the new process for one week. Tell her you will stop by multiple times per day to answer any questions and see how it is going. You have to earn the operator's trust and it will take time. Alternatively, you can kill trust and stop Lean in its tracks by forcing the person to work in the new process because you said to do it! Just kidding, I know you Lean thinkers would never do that.

Following is the attendee survey feedback I received from the event organizer after I returned home from this trip to Australia. Because this event was so successful, which is reflected in the survey responses, it paved the way for multiple trips back to Australia. With each trip, I meet wonderful people who are helping me to change the world, or at least how the world views workplace safety.

Lean Safety Workshop/Kaizen Event Feedback

Impressions—What impressed you the most?

- The way tools can be applied to actual issues
- Potential of kaizen blitz to engage and empower people
- The passion of participants
- I felt very comfortable (not overwhelmed)
- The cross-industry experience, sharing, and networking
- Engagement demonstrated by the site operators
- Use Lean tools to improve safety
- The content of the information provided
- Very good use of practical exercises to reinforce the principles
- The dedication and commitment to change and reducing injuries
- How simple the approach to improving safety can be
- Bob Hafey's presentation style and engagement of class
- Ability of the presenter to keep the team engaged throughout the two days; very good amount of class interaction
- The approach—not production- or time-based—purely about safety and improvement

What can you take away and apply in your workplace?

- The practical use of a kaizen blitz
- The use of Lean tools for safety applications
- The application method
- Kaizen blitzes
- Leader safety standard work

- Engage users of safety systems using Lean tools (process mapping, etc.) to review systems
- Use small teams to blitz areas of concern
- Focus on manual handling
- Root cause analysis
- Quick action for issues
- New means of looking at safety
- Understanding the Lean part of safety
- Kaizen blitz could be applied in areas where ergonomics is an issue
- Application of process mapping
- Use mind mapping with the safety committee to develop policy, procedures, and forms
- Involvement of the people who actually do the task/work, not just asking them about what they do but how would they change it
- Before and after photos
- Kaizen blitz is not difficult—just try it
- Looking for out of neutral body positions
- Ways to engage all levels of the workforce to participate actively in building a positive safety culture
- Leader standard work—all levels of leaders engaged/showing commitment to safety
- Kaizen blitz—an effective way to facilitate change
- The opportunity to learn new techniques such as process mapping
- Need to look at people and the process
- Never, never focus only on the person and what the person did
- Continue to use the 5 whys; this is a really good way of getting to the root cause
- Involve the team members in the solution
- Take time to evaluate—do not rush
- Engagement is key
- Engagement with the workforce
- Refocus on simple issues with simple solutions
- Organization of improvement team
- Application of safety risk squares—measuring the number of times you are in the "yellow zone" as a leading indicator
- Leader safety standard work
- Kaizen safety blitz—looking at the people not the process

Learnings—Personal learnings for me

- Kaizen blitz
- Safety is a continuous improvement process—do I respect you when I let you work every day without growing? I will use this as the way I manage my team
- Ask questions—don't tell. Safety is not hard but it is easy to get wrong. Involve all levels/roles in the evaluation process
- Lean can easily be applied to safety
- Use the Lean tools for safety
- You can get more value through engagement with shop floor operators—richer ideas and ownership
- Ensure you engage the operator in any process change
- Opportunity for reflection on what I am doing at my workplace
- Apply kaizen
- Always involve operators
- Opportunity to facilitate kaizen blitzes

Case Study 9

Recreational Vehicle Plant, United States

"If you don't know where you're going, you'll probably wind up someplace else."

—Yogi Berra

At this site, I only visited a small portion of what was a very large operation. The manufacturing processes required to produce recreational vehicles are labor intensive and ergonomically challenging. The products are low volume and custom, which allows the customer to select a wide variety of options. This makes their final assembly processes comparable to the assembly of a high-end luxury automobile like a Maserati rather than cars in a Ford assembly plant. I have had the opportunity to visit recreational vehicle assembly plants on two different continents and the challenges and opportunities are the same.

While in Australia, I was given a tour of a plant producing recreational vehicles for the domestic market. The customization required for their market certainly made their products look different from those sold in the United States. Because so much of the Australian continent is arid and sparsely populated, users are often driving their recreational vehicles over terrain we would consider an off-road landscape. While traveling around southern Australia, I observed RVs that had metal exoskeletons that added the rigidity required to keep them from shaking themselves apart while traveling in the outback. In addition, the rear panel on some of the RVs was hinged at the bottom so you could lower it to provide a ramp that allowed an all-terrain vehicle or motorcycle to be driven into the unit. All of these unique features reflected the difference in the Australian market.

The plant I toured allowed me to visit all parts of its operation. The area that caused me, as a Lean consultant, to drool because of all of the Lean Safety opportunities was the final assembly area. It was a gold mine because very little automation was utilized, which meant the components were almost all handled during assembly onto and into the chassis. I observed more body parts out of neutral than there are kangaroos in the outback. Unfortunately, I was just a guest on a tour and did not have a chance to affect the operation.

That was different at this site in the United States. My visit was a scheduled half-day event during which I led eight individuals on a Lean Safety Gemba Walk. Our goal was to walk through this very large operation and record as many opportunities as we could in just a few hours. The area we walked was responsible for assembling many of the internal wooden-framed parts used inside of an RV, many of which were later covered with upholstery or a plastic laminate. At our first stop on the tour, we observed associates stapling small pieces of lumber into a variety of geometric shapes. On the carts that held their finished parts were doorframes and other rectangular-shaped parts that would go into the interior of the finished RVs. Items the safety kaizen team recorded on the opportunity log related to table height adjustments, part drawing accessibility and visibility, repetitive motion, awkward positions, and the repetitive stapling action. All agreed that contacting stapler tool manufacturers to discuss new tool options that in the long term may reduce the stress caused by the very repetitive jarring action should be a priority. We moved on past many other work centers and were able to identify numerous soft tissue injury risks that related to material handling heights and positions. One of our final stops on the tour was a work cell where upholstery was applied to wooden parts.

It was here that I observed examples of straining that could be filmed and used as a Lean Safety instructional video. We all know that repetitive tasks can lead to soft tissue injuries. For instance, sitting here typing on my laptop can cause carpal tunnel syndrome if I fail to take breaks and complete some simple yet effective stretching exercises to relieve the strain I feel in my wrists. Typing is a straightforward task that does not include straining to complete the task. Any time a person is required to strain to complete a work task, it increases the possibility of soft tissue injury dramatically. Any upholstery operation requires the fabric to be stretched over the structure or frame being upholstered so that it has a smooth, unwrinkled surface when completed. What we observed were associates using their hands, wrists, and arms to pull, tug, and hold the material in place while it was glued or stapled into its permanent position. When asked if they ever experienced hand, wrist, or arm discomfort, they responded with a yes. We huddled with them and discussed tooling options that might reduce the physical stress required to complete this strenuous work task that has been performed the same way forever. This again is a result of the business model—low volume, high mix—which keeps the engineers and managers from thinking about automating processes like this because they can never justify the investment using a return on investment analysis. Our discussions centered on developing some simple fixtures that utilized quick clamps to hold the fabric in place. One of the downsides of taking Lean Safety Gemba Walks in plants to which I do not return is I have no idea what changes were made. I sincerely hope the internal staff, who accompanied me on this walk, made a safety difference in the upholstery area after my departure.

Case Study 10

Food Plant, United States

"You have brains in your head. You have feet in your shoes. You can steer yourself any direction you choose. You're on your own. And you know what you know. And YOU are the one who'll decide where to go..."

—Dr. Seuss, *Oh, the Places You'll Go!*

The management team at this business decided to explore the Lean philosophy and its applicability to their type of business. If you are a Lean thinker, you understand Lean principles not only apply to manufacturing but also can have a dramatic impact on the internal processes within hospitals, service companies, schools, and government institutions. "A process is a process" is one of the mantras of Lean. Any process can be mapped, taken apart, and put back together in a better way. My initial meeting at this site included most of its leadership team including the CEO. During my initial meeting, I was informed the company went through a very serious financial crisis just a few years prior. This was a red flag for me because it might mean they believed Lean to be a cost-savings program that was going to help turn the business around. Yes, Lean can reduce costs short term, but if that is the goal it must also be recognized that the effort will destroy the relationships and personal bonds that bind the business together. I provided them a brief overview of the Lean philosophy—focus on the customer, engage all stakeholders in waste elimination activities in order to reduce business process cycle times, etc. Since I am a Lean consultant, I do not always start with my Lean Safety philosophy. Business leaders are interested in business results—as in dollars and cents. Using safety as the entry point to build trust and move Lean forward in a business is the discussion I always want to start with but I am perceptive enough to know that conversation can wait until later. Therefore, when I offered to take some of the meeting participants on a Gemba walk to start their "waste identification" education, at the outset I held back on the safety observations that would make work safer and easier. Instead, I focused on and centered my comments on current state WIP inventory, layouts, disorganization, lack of visual communications, and the flow of products as any Lean consultant might do.

Before we could begin our walk into the plant, we had to put on the garments required to maintain their food safety standards—lab coat, head net, gloves, safety glasses, etc. We were issued the appropriate items at their central storeroom location. While standing there, I observed associates waiting in line for gloves. This was symbolic of a facility that had low trust levels. Making adults wait in line for something as simple as a pair of gloves required to complete their work tasks did not value the employees, sent a clear "we do not trust you" message, and was "waiting waste." When I pointed this out, a few of them sheepishly acknowledged that this was a known problem and yet no one was providing the leadership decision making required to initiate the change.

Wrapped in our protective apparel, we began our walk and almost immediately observed a line of pallets containing finished goods sitting in the main pedestrian aisle. I used the opportunity to talk about inventory waste and we continued to walk. In an area that contained a long line of small batch processing units, I noticed a huge wheeled rectangular product bin being pushed toward a packaging station. I was told it weighed about 2,000 pounds and because the floor surface was pitted and uneven, moving this unit was a struggle. They explained the product from the small batch units was dumped into this rolling container and then it was positioned in front of a scale and small carton-sealing machine. A large hoist was used to lift the far end of the large container so that gravity would shift the product toward the packaging station. I restated what I had heard. "You mean you consolidate a bunch of small batches into one big batch so you can break the big batch down into small batches?" I took this opportunity to reference a Lean Safety opportunity. I asked them how the employees who had to push a cart that weighed 2,000 pounds across an uneven pitted floor felt. I explained that if they did an Internet search on "average cost of workers' compensation back injuries" they would note, as I did, that the figure often quoted is $50,000. Therefore, a company that has a 10% margin on their products would need $500,000 in sales to cover the cost of one back injury. I also pointed out that this was "over-processing" waste. Moving products into temporary storage containers when they should go directly into the finished goods cartons was a real cycle-time gain opportunity.

In another production area, a raw material that was purchased in paper bags had to be torn open and dumped into a rolling barrel-type storage container. Invariably this dumping step resulted in product being spilled onto the floor. This container was rolled in front of a line of product processing units. Product was scooped from the barrel and added to the units as required. I pointed out that the container they purchased the product in (a bag) required the physical (safety) and time wasting (cycle time) steps of cutting open and dumping the bag into the container they really wanted— a barrel. The obvious opportunity was to work with the supplier to see if it could ship the product in a barrel-type container that would eliminate both the dumping step and the cleanup of the floor. This observation helped them understand that safety and Lean are inextricably linked.

At the next stop, we observed a purchased product being coated with a melted candy coating. The purchased product was shipped in small cartons that had to be cut open and dumped onto an in-feed conveyor that fed a large-width conveyor on the coating line. It was important that the product did not overlap because that would cause the items to stick together when coated. Here I witnessed a prime example of defect waste. Because occasionally product would overlap, an associate was assigned the role of hunching over the conveyor so she could pick off any overlapped product. I started asking why five times to help direct them to the root cause. Why is she positioned there? What is she doing? Why does she have to do it? Does that work make her feel valued? Why does the process allow product to overlap? What can you do to ensure product never overlaps? If you could do that, what could she do that would add value? A quick huddle around the machine resulted in ideas to use something like silicone food-safe brushes to ensure only a single layer of product advanced on the conveyor. At the discharge end of this line, additional waste was observed. Associates who were directing the finished product into cartons had to lean forward with their backs out of neutral because the platforms on which the cartons rested were too low. They then had to pick up the carton, turn, and walk to a scale where they validated the weight. They again picked up the carton and walked to a pallet where they bent their backs out of neutral to place the carton on the pallet. Transportation and motion waste, plus all of the inherent safety risks, was visible everywhere. I helped them to understand how a new Lean layout that focused on fulfilling relationships, density, velocity, and flow would eliminate the waste we were observing and would make the work tasks easier and safer.

During our walk, they had many side conversations and took notes. Before leaving the site, we once again gathered in the meeting room in which we had started. I asked each of them to share their thoughts about the walk we had just taken. The responses indicated that some of their eyes had been opened and they were anxious to pursue some of the opportunities we had identified. Others—not so much. Over the next couple of months, I returned to this facility to provide some Lean 101 basic training for the same group and finally to help them value stream map their upper level customer delivery business processes. One common element I recognized on all of my visits was the lack of Lean leadership at the senior manager level.

In the beginning of a Lean journey, the senior leader of a business has to push Lean into the faces of his or her senior staff. The leader has to make them uncomfortable by displaying an air of healthy discontent. What do I mean by that? He or she has to challenge the current processes and principles that each of his or her senior staff members are trying to hold onto. Each of them is responsible for a functional area of the business. A leader who is truly a Lean thinker understands all functional areas must reassess how they support the new business objectives by utilizing Lean thinking and tools. When one or more of the senior leaders dig in their heels or drag their feet on the Lean journey, not only can it slow progress but also kill the program. The most senior leader must visibly challenge the status quo, the old way of doing business, and any individual in a senior position who is not supportive. In too many businesses, including this one, senior leaders think the middle managers, who have shown some passion for Lean, can move the program forward. Lean is not a program; it is a trust building exercise that requires you to take apart and rebuild a business. Only the most senior leaders have the authority to demand acceptance or departure from those who do not show support. If they do neither, two factions will develop—one group in favor of preserving the status quo and the second one anxious for real change. Lean cannot be led from the middle. It has to be led from the top—period. Leaders who do not lead the change effort are not serious about Lean—they are not Lean thinkers. That visible lack of passion frustrates everyone in the organization who has seen the value of Lean and would like to move the effort forward. The company ends up in a state of confusion simply caused by a lack of real Lean leadership. Lean dies a slow death and the company culture returns to one of finger pointing and firefighting where continuously coping is deemed more valuable than continuously improving.

Case Study 11

Sheet Metal Fabrication Plant
KSO Metalfab Inc., Streamwood, Illinois

"We have had many trainers come to our company throughout the years but never have we had someone that cared the way you do. You have made a difference in everyone that was in the training not only here at work, but in their (personal) life."

Up until 3½ years ago, I had always worked for someone else. "Working for the man" had its pluses and minuses. The most obvious benefit is a steady income, which is important when you are raising a family and paying off a home mortgage. On the downside are a limited, or narrow, opportunity for personal growth plus the many common elements of a corporate culture, like the numerous and sometimes seemingly endless stream of meetings, that gobble up your time. In addition, there are all of the HR rules, regulations, and forms one has to comply with and complete. Finally, there are the reporting relationships and the company politics that are part of every job when you are working for the man. Then at the start of 2010, I became the man. I formed my own limited liability company. I am the only employee. I do the sales and marketing, systems work, accounting, material development, travel planning, workshop presentations, keynotes, and anything else the business requires. It was liberating and unbelievably exciting not only to set my own path but also to be successful on the journey. I truly am having the time of my life. Then in mid-2013, I received a call from "the man."

The man, in this case, was a consulting group. After finding me on the web, they called to ask if I could visit their office for an introductory meeting. After two meetings, I was offered a part-time position as a Lean consultant. To be honest, I was reluctant to accept the position because I really enjoyed my independence. Working for yourself will always beat working for the man. Finally, I said I would try it. Our working arrangement was based on a handshake agreement and I was assured there would be no meetings to attend. Not bad, I thought! Then I received an email asking me to log into the company database to complete the required employment forms. After a few frustrating attempts, I was questioning the wisdom of agreeing to work for the man again. Yet something made me persevere. That was the promise of new opportunities to engage people and help them believe in themselves. So, I thought of my new employer as an agent who preselects clients who I could help on their Lean and safety journeys.

One of the client companies I was asked to visit was a small sheet metal fabrication firm with just over 50 employees. The current leaders, I was told, were very interested in changing the culture of the business using the Lean philosophy. On their own, they had started with a 5S effort that resulted in limited success. They were frustrated with the lack of progress and had called the consulting group asking for help. Therefore, I called one of the owners and scheduled a visit. I was cautiously greeted and after sharing our backgrounds with each other, I asked my host if we could take a Gemba walk to help me better understand the business, the opportunities for improvement, and the workplace culture. I knew I had to earn his trust and the best place for that is not sitting in an office but on the Gemba. It is there where I could demonstrate the value I could bring to his business. So together we walked the flow from raw material receipt to the shipping area where finished goods were packaged for delivery to customers. I observed WIP inventory sitting in front of almost all the work centers (inventory waste), associates grinding the edges of sheet metal parts because the cutting operation had left burrs (defect waste), a plant layout that interrupted the flow of parts (transportation waste), and component inventory stored on a mezzanine (transportation waste) rather than in the work centers where the parts were consumed. In the final assembly area, associates were walking back and forth (motion waste) and were observed laying and kneeling on the floor to assemble finished parts. As we walked, I not only observed and asked questions but also was trying to determine what training I could provide that would have an impact on the business. The only constraint I had was that the training had to be conducted in the next five weeks.

After we completed our walk, we sat down to develop a training plan. Because I was already scheduled for some travel, including a weeklong trip to Texas to conduct a Lean Safety workshop, my availability was limited to about three weeks. Based on the time available, we developed a training plan schedule that totaled nine days. A lot of the initial Lean training that occurs in most businesses is conducted in a classroom. Spending multiple days reviewing PowerPoint slides on 5S, SMED, the seven wastes, kaizen blitz events, value stream mapping, and the many other Lean tools is easy consulting work. That is not how I like to spend my time. I like to take trainees to the Gemba and allow them to learn by doing. People who work in manufacturing plants like this one do not like to sit in meetings or training classes. They are hands-on learners and I wanted to provide them the opportunity to make a tangible difference in the business. Over the nine days, the trainees would be exposed to 5S, set-up reduction, plant/cell layout, process mapping, workflow improvement, kaizen events, and, of course, the opportunity to make work safer and easier for themselves and their co-workers. Moreover, when I say "exposed to" I do not mean in a classroom. I mean on the Gemba.

If you have not been in a manufacturing facility in the last 10 years, you may not know that Hispanics, just as they are in the U.S. population, are becoming the majority. I often reflect on the fact that we are all children, or descendants, of immigrants. Each of us can search our family tree and be transported back to a time when our ancestors left some distant land and came to America for the promise of a better life. Many of them, just like the Hispanics today, worked in factories. Because language is an issue for all immigrants, they prove their worth by working very hard. As the years pass and both their language skills and work knowledge grow, they are often given leadership opportunities. Because they have always worked hard and have not been trained to lead, they often accept the promotion but then feel uncomfortable in their new role. This describes some of the trainees with whom I worked over the nine training days.

On Day 1, they all appeared to be a little nervous and unsure about what was in store for them. I asked each of them to introduce themselves, tell me about their role at work, and one passion they had outside of work. It was the beginning of a process intended to build a core Lean team that believed in themselves and would impact the business after I departed. This initial meeting started in a classroom with a book on 5S for operators in front of each of them. I asked them to turn to and then read the two pages that gave the definition for a 5S program and a brief description of each "S." I asked if they had any questions. If not, we were finished in the classroom because it was time for them to guide me on a Gemba walk of their plant.

I asked them to guide me following the product flow and while doing so I asked them to think about which department or work center we were going to apply the principles of 5S. After we had finished the tour, I asked them which would be the best area to apply 5S. To help them understand how to make those choices I then asked, "Which department is the one in which no one wants to work?" I asked the question at this time because we were standing outside the sanding/deburring room. It is where the sheet metal parts were sanded and had their edges deburred. The entry level and temporary associates who worked here stood hunched over parts with their backs out of neutral in a noisy, dirty, and smelly environment. The walls, floors, and equipment were dirty and covered with grinding dust despite the recently installed new dust collection system. When they hesitated in their response to my question, I asked, "Which one of you wants to work in there?" I pointed through the dangling plastic strip curtain intended to keep the dirt and dust contained in the room. They agreed; no one does! Then I asked if they wanted to help make it a better workplace—cleaner, brighter, organized, and safer. For the balance of the day, working in small sub-teams, they implemented the 5Ss—sort, set in order, shine, standardize, and sustain. Floors, walls, and equipment were cleaned, tooling was organized, and benches and equipment were relocated to support an improved product flow. That night after we departed, the site manager arranged to have the room's walls painted white with a decorative blue strip at waist height. The next morning, the teams continued by organizing and labeling tooling and equipment. A kanban-replenished mini-storeroom was set up in the grinding room to eliminate the walking waste required to obtain grinding discs, gloves, and respirator masks from the current central stores location. Meanwhile, one of the trainees developed the standardized plan to ensure that the first three "Ss" were maintained. This was immersion learning at its best. In just a day and a half, they learned how to implement 5S and they understood the benefits of doing so because it was reflected on the faces of the associates who worked in the grinding room —big smiles and nods of approval. They now worked in a place in which they could take a little pride. The trainees left for home at the end of the day feeling good about what they had accomplished and returned the next morning anxious to hear what we would do next.

This was a make-to-order job shop but they had some customers who relied on them to build and deliver a steady stream of the same products. Knowing this, I gave them an overview of process mapping and then asked them to process map the steps for one of these repetitive products. This exercise was to help the trainees understand flow or, in their case, the lack of flow caused by large batch sizes and their departmentalized plant layout. As soon as the process map was completed, I asked them to take me on a Gemba walk so we could walk the flow of these same products in order to challenge the current layout. Every time a batch was put on a pallet or cart and moved across the plant to the next work center, I pointed out the waste inherent in their layout. Once they understood how important a facility's layout is to good flow, we returned to the classroom.

I provided them an overview of block and relationship diagramming, which is used to define new plant layouts that support flow of product to customer. They defined the blocks, cut out pieces of paper to represent the blocks, and defined a new plant layout that would fulfill the equipment relationships. Even with a great layout, batch production slows the flow. To help them understand the ill effects of batch production on product flow, we went back to the Gemba. As we stood in front of batches of components waiting to be processed, I asked them to calculate the total cycle time to process each batch. We did the same for all of the process steps and batches that went into a repetitive product so we could estimate the total cycle time to customer. To emphasize and really drive this point home, we headed back to the classroom. I led them through a simulation in which they built paper airplanes. They started by building batches of five with a disconnected layout. After two other iterations of the same exercise with work-center layout and batch quantity changes, we arrived at the final stage—a one-piece flow pull signal production process. In each timed cycle of this simulation, they measured output, defects, and WIP inventory. This simple exercise really opened their eyes to the prime benefit of improved layouts and smaller batch sizes—reduced cycle time to customer. By now, we had completed about five of our scheduled nine training days and they were each immersed in their own personal Lean journey that challenged not only the work processes but also each of them individually.

The next stop on their Lean learning journey was to participate on a kaizen blitz team charged with reducing the changeover cycle time on a piece of metal-forming equipment. They learned about teamwork, challenging the status quo, direct observation to identify waste and opportunities for improvement, doing rather than talking about it, and defining the best business process (standard work). As they watched the flow of material coming to and leaving the targeted work center, they began to question the layout and the batch sizes. They were trying to improve flow by linked work centers and reducing batch sizes. That made me very happy!

Their final project was to tackle a final assembly work center and, using all the Lean tools, improve the flow to the customer of a large electrical cabinet. The current state assembly process was an easy target because the assemblers followed no standard work processes. They worked on batches, had to walk for components and tools, and had some body part out of neutral more often than not. After observing the current state assembly process and recording the opportunities on their log, the team set to work defining a one-piece flow assembly process. This required them to define the steps of the assembly process and then design a layout that would support that defined flow. They also designed and built an assembly bench with a hinged sloping end onto which the assemblers could position and easily remove the cabinet. Components that were previously assembled inside the cabinet were now assembled on a workbench and then inserted into the cabinet as one complete unit. Components and tools required during assembly were located on a rolling tool board and positioned at point of use. The team was on a roll applying all they had learned in the previous seven training days and I was able to stand back and watch both their abilities and confidence grow. On the ninth and final day of training, they were going to observe their redesigned cabinet assembly process to fully understand the impact on the work process and the safety of the assemblers. This final project had required them to apply all the lessons learned in their earlier training— process mapping, layout design, 5S, reducing cycle times, one-piece flow, and focusing on the safety of their associates during the new assemble cell design and build. Then when the first cabinet was completed, they, as all teams who have a common focus and are successful, were beaming with pride at their collective success.

We then gathered back in the training room to debrief and begin to prepare for a presentation to management, which would take place after lunch. Each of the trainees, as a lead person, supervised and directed others. They were selected for this Lean training because they were being asked by management to take on more leadership responsibilities, which made them very uncomfortable because they worked in a culture where they had always been told what to do. This is not unusual in small, privately held businesses. Only leadership can solve this leadership problem. It is widely understood that if you continually give people the answers, you take away their responsibility to solve the problems that occur in the business. Therefore, managers who have always relied on a top-down directive style of leadership must change their behaviors if they expect to build trust and grow their reports. They have to adopt a coaching style of leadership that results in them asking the right questions rather than having the right answers. All of the trainees had concerns related to their ability to lead Lean in their business because of the current culture, so over our last lunch together they shared their feelings with me—someone they trusted because I had respectfully pushed them to learn and grow.

The report out to management was a real growth opportunity for all of them. Rarely if ever had they been asked to get up in front of a group of people and present. The reward for Lean leadership, whether you lead Lean in your business or someone else's as a consultant, is watching the people you have touched grow. I felt like I had won the lottery as I watched them report on their training and the results of their efforts. Even better was watching the expressions on the business leaders' faces as they beamed with pride while watching their staff.

For nine days, these trainees had been engaged. They had been given the gift of time to focus on continuous improvement. The word "engaged" is often bantered around businesses. To clarify its meaning in my training sessions, I approach the topic of employee engagement by asking the attendees to tell me about their "best day at work." I believe people find satisfaction in the work they perform and this question allows me to explore that topic with them. Everyone will agree that some days are better than others, but what I want to explore are the elements of the best day. Answers I get to the question vary, but the underlying theme remains the same. Some common responses are:

■ I was productive.
■ My equipment ran great all day.
■ I had everything I needed to do my job that day.
■ I was left alone to do my job.

Based on these responses, people want to come to work and do a good job and, as I stated previously, do find satisfaction in the work they perform. The problem with this scenario is that many people expect their job to remain the same. In this global economy, doing anything the same way, day after day, will lead to mediocrity and business decline. In a continuous improvement or Lean business culture, leadership expects constant change—change for the better. It is only by letting go of control, trusting that all employees can handle more, and then driving decision making downward within the organization that a management team can truly engage their workforce in ongoing business process change. When this occurs, it gives many more employees the opportunity to walk out the front door at that end of their shift feeling like they made a difference that day. That is the common theme echoed in the bullet pointed responses above—anyone's best day at work, and it matters not if they are hourly or management, is a day when they personally feel, in their heart of hearts, like they made a difference. My trainees felt exactly that way. The big question that remained as I was walking out the door was would they continue to feel that way? Only if management started asking the right questions rather than giving the right answers.

Case Study 12

Medical Device Component Plant
Specialty Silicone Fabricators, Paso Robles, California

"Robert presented a Lean Safety seminar at our facility in Paso Robles. He did an excellent job of getting people thinking about improvement opportunities and providing a path forward to implement them. Our staff came out of the seminar very engaged and motivated to implement positive change."

This trip took me to Paso Robles, California. If you are not familiar with its location, it is about a 3-hour drive south of San Francisco. This is the central valley area of California best known for agricultural products. It was my first time in this area of the United States and I really enjoyed my stay. There was plenty to do and see with the California coast no more than a 30-minute drive west. The recently retired president of this company was a Lean pioneer I had met through the AME organization some years ago. We were both serving on the AME national board when I became aware that he had created one of the first Lean-centric websites called Super Factory. I also remember reading about Paso Robles in some of his later blog posts and tweets. Like me, he enjoyed a fine meal accompanied by a glass of good red wine, so his occasional description of dining in and around Paso Robles had created a bit of an aura in my mind. I had been invited to this site by the Western AME board. They promoted the event and filled the seats for this two-day public Lean Safety workshop hosted by a medical device component manufacturer. The company, located in a brand new facility, was staffed with friendly, knowledgeable, and professional people. The senior level contact for the site had been delayed in his return from a business trip. When I heard that he was not there I was not alarmed at all because this was a company with a Lean culture in place. I was very confident others would step forward to fill the gap. That is exactly what happened during the preparation and start of the event when the HR staff and others helped with the set-up of the room, AV equipment, and refreshments. I asked one of the other internal contacts, who had stepped forward to assist, if he could think about the work center assignments for the small kaizen teams that would work on the Gemba on Day 2. During morning break, he talked with a couple of the supervisors and then returned to ask if I could meet with all of the supervisors. We arranged a meeting time during the afternoon break.

This communication step is critical when planning events that are held in a work area. Everyone wants to know what is going on when a large group of visitors is in the facility. This is especially true if they will be visiting their area of responsibility. No one likes to be surprised. When we finally had our meeting, I explained who I was and why I had been invited to visit their facility. I gave them an overview of the Lean Safety concepts and the value the concepts could bring to their business. I asked them for their help in identifying work areas or individual work tasks where they had ergonomic or safety concerns. Together we set about defining the areas where the kaizen teams would be assigned because they felt the teams would have a positive safety impact there. We finished our meeting on the same page and in agreement on the expected outcomes.

By now, you are all familiar with the format of a two-day Lean Safety workshop—PowerPoint slides and small team exercises concluding with a Safety Gemba Walk through the plant on Day 1. Before taking the Safety Gemba Walk, we all had to don the protective gear required to enter a clean room environment. As is usually the case, a group of the workshop attendees were from the host site. A few of them acted as our tour guides as we began the walk. The work processes we observed varied from physically demanding to visually demanding. At one of the first work centers, a machine with revolving horizontal rolls mixed a thick compound. The operator had to physically grab the leading edge from the discharge side of the rolls and insert the large lump of product back into the top of the rolls— think of a pasta machine only a lot larger. It was a risky operation because the rolls turned constantly and the operator's hands came close to the roll entry point. One of our tour guides worked in the area so he demonstrated the job task and pointed out the two emergency stops that the operator could activate if necessary. It still made me nervous and I would experiment with vertical rolls if I worked at this site.

In another work center, we observed two operators moving molds in and out of heat presses. Sounds like a dangerous task, but the parts produced in these molds were so small that an associate who was deburring the parts after they were removed from the mold was using a microscope to view them! The opportunities identified included the workstation layout and work surface heights. Adjusting both would help to keep the associate's arms, shoulders, and back in neutral. I think that many individuals feel the larger and heavier the parts, the more the safety risk. That might be true if you have to handle and move heavy parts. You can be pinched, crushed, and injured seriously when working around very heavy large parts. Yet soft tissue injuries can occur when handling the smallest of parts if the work is repetitive and you are not positioned correctly. The injury type that taught us all about ergonomics and office workstation layout was carpal tunnel syndrome. The simple task of typing caused many people to suffer from carpal tunnel syndrome.

I was in a medical device assembly facility outside of Sydney, Australia a few years ago. My contact in Sydney had arranged this visit based on the site manager's interest in meeting me and learning more about Lean Safety. In the initial meeting in which I presented a brief overview of Lean Safety, the consensus feedback from the site staff was that they really did not have any safety issues. "You know," they said, "small parts and light assembly work is all we do here." I smiled and asked if we could take a Lean Safety Gemba Walk. They agreed and I quickly opened their eyes to all of the safety risks that existed day in and day out in their workplace. A prime example was a person who was inserting components into a cabinet. She and her co-workers would add components much like the assembly process in an auto assembly plant. The cabinet had wheels and was rolled from one work center to the next. When we observed her working, she was bent over the cabinet installing a part and hooking up the electrical leads to the appropriate connections. I walked up to her and asked how she was doing or, as they say in Australia, going. She said, "Fine." I said, "I noticed that you have to continually bend over to install the components into and onto the cabinet. Does you back ever hurt?" After she said yes, we talked about raising the cabinet by putting it on a rolling dolly to help keep her back in a neutral position.

At a plant in San Antonio, Texas, an associate was sitting in front of a magnifier working on a small part in a fixture that was permanently mounted to the workbench. Because the fixture was mounted too far forward on the bench, the associate had to work bent forward with his back out of neutral. I asked him if his back hurt. "Yes," he said. "What if you moved the fixture back toward you? Would that help?" I asked. "I think so," he said. The perception that small, light work is less risky often results in poor workstation designs. The industrial engineers spend their time where they believe the risk is greater. What I have learned is that you do not have to have injuries to signal an unsafe workplace. You may have zero recordable injuries, yet people go home sore and take painkillers every night. The workplace is unsafe but the current compliance-driven safety metrics do not reflect that reality. Of course, you could track the associates' reported use of painkillers and work together to drive the number down! Lean Safety Gemba Walks identify and eliminate those hidden risks. Take a walk and see for yourself.

Case Study 13

Coiled Metal Processing Plant, Australia

"Thank you for the opportunity to attend the Lean Safety Conference conducted by Robert Hafey. This was a real eye opener for it has definitely changed my mind set, scope, and further potential interactions with staff. This course highlighted and provided the tools for 'Going to the Gemba and Kaizen Blitzing' and gave us all the fundamental basics and understanding of how we can improve at a site level. This course has solidified our potential and current systems of improvement, recently incorporated by our internal continuous improvement program. Our objective on Day 2 was to simply observe the workplace practices and engage the staff—with our focus on ergonomics. This, from my perspective, gave us the key formula for cultural change, increased productivity, and formulating safe work practices. Employee engagement + ergonomic improvement = reduction in cycle time + increased production + positive culture. As in any change or questioning of current practises, we were met with resistance, yet this was overcome by explaining the 'method of our madness' and how we were there to make their particular jobs easier. Once the barriers were broken down, it became easier to assess and put forward the opportunities for improvement."

I love it when workshop attendees really take the Lean Safety training to heart and think about how they can use their new skill set to make a safety difference when they return to work. The quote above exemplifies that type

113

of an attendee. He forwarded the above message to his supervisor, who had invited him to attend the workshop, the morning after the two-day event. The supervisor forwarded the message to me and it made my day. My Australian contact, after reading it, noted, "This is gold!" What he meant was I would be able to use this to market the workshop when I return to Australia next year. Just as all other businesses, my success as an independent consultant is based solely on my ability to impact people and the businesses in which they work. There is no organizational corporate culture or image that I can use or hide behind—it is just me developing one-on-one relationships with people all over the world that defines my success as a consultant. In some people's eyes, placing the title "consultant" after your name tarnishes it. I have noticed a trend recently that has some former consultants now calling themselves "executive coaches" or some other variation or combination of words in an attempt to distance them from consulting. I am okay with being a consultant and think the people I meet and impact see me as a resource with a new meaningful message that is centered on helping people. Words are just words—our actions speak to who we really are.

This site, a member of a consortium of companies, hosted a two-day Lean Safety workshop. They processed coils of steel in a very large and impressive operation near Brisbane, Australia. Our hosts led us on the Day 1 ending Lean Safety Gemba Walk. As the facilitator, I understand that these initial walks through a facility serve a dual purpose. First, I must allow the internal staff the opportunity to showcase their operation. They are proud of what they do and how they do it. The outside attendees are also very interested in learning about the operation. I am a big fan of this "industrial tourism." I have been in plants all over the world and I never tire of those learning experiences. I have seen unique processes from croissants being made (what a smell!) to watching molten steel cast into huge slabs, and learned from each. The common ingredient in every business process is the people. Therefore, at some point on each plant tour I must focus them on the second reason for our walk together—to observe people at work. Simple questions like, "what did you just see that person doing?" can take them from tour mode into the Lean Safety observer mode. At first, many of them are not sure what I mean and I know what they are thinking. He was working! Early on during our walk, we observed an associate banding a coil of steel, positioned on a mandrel, with steel banding or strapping. What they observed was the associate placing five bands around the coil and then using a mechanical bander to tighten them to prevent the coil from uncoiling. He then moved outside of an interlocked gate and moved a motorized cart under the coil. The mandrel was retracted so the coil rested on the cart for removal. The cart was moved out of the mandrel area so a hoist could be used to take the coil to an inventory location.

Because I challenged them to see it differently, before we departed from the area, what they initially observed had changed. They now observed the associate walk back and forth six times to retrieve a precut length of banding and then get into very awkward out of neutral positions to place the banding around the coil. The area he was working in was compact and tight. It was inherently unsafe because it had an interlocked gate access. He then had to physically lift and hold the mechanical bander into position as it tightened each of the six straps. He then moved outside of an interlocked gate and moved a motorized cart under the coil. The mandrel was retracted so the coil rested on the cart for removal. The cart was moved out of the mandrel area so a hoist could be used to take the coil to an inventory location. Because they went beyond watching work happening to closely watching the individual performing the work, some great conversations about making work safer and easier took place. Some of the ideas were:

■ Contain the coil with a single nylon strap, tape, or clamping device while it is on the mandrel
■ Move the coil onto the cart and out of the mandrel area so it could be strapped for shipment while sitting on the cart. This would keep the associate in a neutral position while banding the coil
■ Position the cut to length banding at point of use
■ Attach the mechanical bander to an overhead balancer that would carry the weight while it is in use

As noted in the quote that began this case study, the associates that are being observed are often a little defensive when you first begin to challenge how the work is currently performed. It is a natural reaction that is overcome when they realize the people making the recommendations are sincerely concerned about their well-being. That is one of the key differences between Lean Safety focused improvement events and traditional Lean kaizen events that focus on cycle-time reduction. We engaged the operators in discussions about their well-being, not cycle-time improvement. Yet if the coils were strapped outside the mandrel area as suggested, the equipment could be started 8 to 10 minutes sooner, which would result in many more coils being processed each shift. That is the beauty of Lean Safety. You make work safer and easier and you reduce cycle times as a result. It is an outcome of your safety improvement activity. That subtle difference allows a business to engage its employees.

On Day 2, the attendees were assigned specific areas to observe and improve. I bounced between the teams to ask the questions required to assess their understanding and move their efforts forward. One of the teams was about to observe a changeover on a coating line. Strip steel was uncoiled a one end of the line, coated, and then recoiled at the other end. The coil line ran continuously with the trailing end of a coil attached to the leading edge of the next coil. Therefore, the changeover of coatings was performed on the fly. A team of two associates was staged like an F1 pit crew ready to perform the changeover at a predetermined time. I have been a fan of F1, an open wheel racing series that originated in Western Europe and now sponsors races around the globe, for many years. Over that time, I have watched the cycle time to complete a pit stop drop as they implemented numerous process improvements. Just last weekend at a race in Austin, Texas, one of the teams changed all four tires on a car in less than 2 seconds. Changeover cycle-time improvement events, often referred to as SMED (single-minute exchange of dies) events, are a very common Lean-driven activity. A common object for a kaizen team is a 50% reduction in the set-up or changeover time. This coating line changeover was different because the process did not stop for the changeover. An F1 equivalent would be trying to change the car's tires as it was driving by at race speed!

This process, like the coil banding process described previously, contained very large equipment. In Lean-speak, these are often referred to as monuments—very large, fixed objects that you are not going to move during a kaizen event. Therefore, you have to accept that fact and find the improvements that work in and around and support that existing layout. Too often people will use the "monument defense" as a reason to justify inaction. It is just one of the many defenses used by those trying to preserve the current way of doing business. Some others are:

- The layout is fixed.
- Management won't support us.
- We didn't budget for it.
- We are a union shop.
- Our culture won't support it.
- We are too busy.
- Lean doesn't apply to our business—we are unique!

None of these is an acceptable excuse today. They are not used because people fear change. What they fear is the unknown impact of the change on themselves. That was not the case during this changeover. We had introduced ourselves to the two associates performing the changeover and explained our objective—to make their work safer and easier. At the appointed time, they donned the appropriate safety apparel and performed the changeover in a coordinated effort.

The task sequence the operators performed was:

1. Open the coating rolls using the appropriate controls
2. Remove a small support roll
3. Clean the current coating from large applicator roll and multiple chutes that are used to channel the excess coating back into a drum

The work environment required the use of respirators because of the coating and the liberal use of solvents to clean the rolls and chutes. The safety kaizen team could not communicate with the associates during the changeover because of the environmental hazards posed by the solvents and coating. We observed through a window in an observation area. What was observed was the continuous and liberal use of solvents to clean everything that touched the coating. It was running and dripping off everything including the associates. To me, this was a prime example of "this is the way we have always done it." People tend to think solvents are the only thing that will effectively clean metal parts if that is all they have ever used. During my career in manufacturing, I worked around some seriously dangerous chemicals that businesses thought they could not do business without—metal cleaning solvents being one of them. Now there is a whole industry that supplies parts washers and soaps to clean metal parts in place of solvents. The case to eliminate or at least minimize the use of solvents during this changeover had not yet been made, but I hope the kaizen team observations caused them to pause and consider the alternatives. After the changeover, the ideas for making the work safer and easier included:

■ Relocating a small solvent tank used to clean the small roll that was removed to a location closer to the point of use
■ Adding dam-type gates to direct coating in drain chutes rather than using bunches of rags to block the flow
■ Using custom form-fitting rubber squeegees to clean chutes and rolls rather than solvent-soaked rags

- Scheduling a kaizen blitz event with the objective of eliminating or reducing the usage of solvents during coating changeovers

We reviewed our findings with the associates after they finished the changeover. I saw the look of "we can't do that ourselves; management has to make those decisions." Yes, in work environments where employees have not been fully engaged and empowered, change waits for management. This business, like many others, was on the Lean journey that never ends. Yet, they were certainly open to change because they hosted the workshop. I can report that the site leadership listened attentively and thanked all the workshop participants for their help in identifying safety improvements during the management report session held at the end of Day 2. While I moved onto Sydney to facilitate another workshop, I hoped they were scheduling a kaizen event to minimize the use of solvent.

Case Study 14

Medical Products Plant, Australia

"A pessimist sees the difficulty in every opportunity; an optimist sees the opportunity in every difficulty."

—Winston Churchill

On a previous trip to Australia, a safety professional from this firm attended my Lean Safety workshop. As often happens, their attendance generated enough interest and passion for the topic that a decision was made to host a workshop at their site. This site was part of a large global medical device company with thousands of employees. It was a beautiful facility with a natural setting including ponds and native plants in between the office and production facility. The corporate mission statement was purposeful and very visible around the facility. I have noticed a trend of corporations developing new mission statements that are purposeful. In the past, most businesses had mission statements that promised quality products and services and the timely delivery of the same to their customers. No need to make those promises to customers anymore because they are expected, and if a company cannot provide them someone else, somewhere in this global economy, will. Instead, newly defined mission statements are being used to engage the company's employees in something bigger than themselves—something that spells out the purpose of the business. As an example, one of my family members works for Ernst & Young, a large public accounting firm. The firm recently went through the rebranding process to develop a new logo and mission. They are now simply called EY and their mission statement reads, "Building a Better Working World." Of course, the rebranding effort at any company comes with a cost—new everything (e.g., business cards, etc.) that contained the old logo and mission. Yet, it is a trend that supports the Lean movement because it is intended to align everyone in a business toward the new purposeful mission. My own personal mission is to change the world—or at least how the world views workplace safety. That gives me focus and aligns my efforts, like writing this book.

This was a well-attended workshop with over 30 people in attendance—about half external and half site employees. The host site had good local representation and had flown in attendees from other locations in Australia and New Zealand. Size matters when facilitating workshops. I limit the attendance number to 30 and when the workshop sells out it is a better experience for both the attendees and me. My energy level along with the energy level in the room seems to mirror the attendance level. I also believe the opportunity to learn improves because they learn from each other as the small teams share the results of their group exercises. For instance, one of the seven exercises I use to involve the small groups (usually teams of 5 to 6 people) is to develop a six-question survey to gauge the level of engagement of their employees in their "current state" safety program. If there are only 12 people or two groups of 6 in the workshop, a total of 12 questions are shared. With the 30 attendees in this workshop, they would hear 30 to 36 possible questions they could select from to take back and use at their site. On the feedback surveys completed by the attendees, at the end of public workshops like this one, they always note the interaction with others as one of the highlights of the workshop.

One of the downsides of filling up a workshop is that when it is time to take the Lean Safety Gemba Walk at the end of Day 1, it is almost impossible to communicate effectively with 30 people as we take the tour. Often, the host site will provide multiple tour guides for this initial walk through and then I have to divide my time so I can spend a part of the tour with each group. A consortium facilitator from Northern Iowa Community College (NIACC) set up the best possible scenario. She arranged headsets for everyone and mine had a microphone so as I engaged the associates in discussion they could all hear what I said and how I asked questions.

This was a big batch production facility. The batch size was driven by a sterilization unit (call it a monument) that was loaded to the max before the long cycle time required was started. This meant that all over the facility the associates were handling batches that they had to put onto the racks used in the sterilization unit and then someone else had to remove them from the racks. None of the racks had been built with the associates or ergonomics in mind. Some required the individual to bend almost to floor level to load the bottom shelf and others were so deep the associate had to go out of neutral to retrieve the product from the back of the shelves. All of this repetitive material handling was driven by that one monument. My guess is some automation projects are in the capital budget at this facility. If they could develop an inline sterilization unit and automate much of the material handling, the flow of products through this facility would improve dramatically. We moved on to observe another operation, which was a finished goods packaging machine.

Once again, we observed an individual unloading a large batch of product from a storage container to feed a conveyor that guided the product through a quality check before it was manually pulled from the conveyor and inserted into corrugated cartons for shipment. At the packaging station, the product was delivered on the conveyor, which was about waist height, and the corrugated cartons into which the product was packed were sitting in between the conveyor and the associates. The team immediately noticed the associates had to bend forward, with backs out of neutral, to retrieve the product and pull it toward them and the cartons into which they would be inserted. I noticed that there were deflectors located at the rear of the conveyor that, if adjusted toward the associates, could drive the product toward the center or near side of the conveyor and would minimize or eliminate the requirement to lean forward to retrieve the product. A short while later during the morning break period, I approached the supervisor of the area and asked if we could have maintenance adjust the deflectors. We agreed to meet at the machine after break so I could show him what the team wanted to do and why it would benefit the associates. When he arrived, he said they couldn't move the deflectors because the product would hang up under the deflector and cause a jam up of product. In fact, he noted it happens occasionally with the deflectors in their current position. I went into the "asking why five times" mode. Why does the product hang up on the deflectors? There is a gap between the deflector and the conveyor surface. Why is there a gap? The deflector is adjusted too high. Why can't you adjust the deflector down to eliminate the gap? Maybe we can but the product might still hang up. Why can't you use a brush of another soft surface on the bottom of the deflector so the product cannot hang up? I never thought of that! Very quickly, the maintenance staff was adjusting the deflectors in and down to eliminate the gap, to test the impact on the associates who were packaging the product. The supervisor offered to follow-up and check into the brush-type deflectors. Both the packaging associates, who had admitted their backs hurt, and the kaizen team members were smiling as the kaizen team walked away. Without stopping and watching people work, this type of opportunity to make work safer and easier goes unnoticed. An experience like this is the essence of Lean Safety and all of the participants will think differently going forward. I often say the biggest value of a kaizen activity is not the process you improve; it is the minds you change.

I think working in a facility that produces products used by individuals who are being treated for a medical condition creates a unique work culture. I could feel it while in this facility. Everyone there understood the need for and accepted a higher level of responsibility for the quality of the products they delivered to those in need. That should be comforting to all of us who might use the products they produce.

Heavy Machined Products Plant, United States

"I would recommend this workshop for others because it teaches engagement and doing versus hoping, talking, and planning."

"What I appreciate about this methodology and approach is the required engagement of the team members. This is not an observation audit nor was it the typical compliance inspection. This process leverages the dialog between the team members and the auditor driving safety improvements and eliminating waste in the process."

Who are the usual attendees sent to workshops and conferences? My observation is that they are managers and professional staff. Each year I attend the largest Lean conference in the world. The Association for Manufacturing Excellence sponsors it and each year approximately 2,000 people are in attendance. The associates who work with their hands to make the products in the companies represented at this conference are a small fraction of the total attendance number. Why is it that companies who are on the Lean journey, and understand that they have to engage their employees to realize Lean success, do not send the people who work with their hands to these events? If you want everyone to think like the managers of the business, then I believe you need to expose everyone, or at least some of them, to the same experiences as the managers. This simple act can accelerate the Lean journey for any business. At this site, I had the opportunity to interact with and influence associates who were United Auto Workers (UAW) members. I mentioned this company earlier in this book. You may remember me describing the manager who wanted a workshop titled, "Coaching for Safety." Then during our meeting to discuss the event, he informed me he was retiring in two weeks! Well, he introduced me to another member of the operational staff who worked with me to plan the two-day workshop I will describe.

The format of this workshop, as well as the attendees who participated, was different from my standard two-day workshop offering. The differences were my two days spent on site were intentionally two weeks apart and the participants were from all levels and areas of the business. The first day of the workshop followed my standard format—PowerPoint slides and small group exercises followed by an extended Lean Safety Gemba Walk. As I was setting up my laptop and other materials for the workshop, the attendees trickled into the meeting room. A few of them, I noticed, were wearing UAW union emblazoned T-shirts. I wear T-shirts and the ones I buy are selected because of the message they intentionally deliver. When I visited the Ukraine last year, the hotel at which I stayed was directly across the street from the largest and newest football (soccer) stadium, Donbass Arena, in Eastern Europe. It was located in the city of Donetsk and was the home of FC (football club) Shakhtar Donetsk. During some down time, I walked around the stadium and the surrounding grounds, which were adjacent to a beautiful park that contained a massive World War II memorial. As I circled the stadium, I noticed a gift shop and walked in to warm up (it was November) and look at the merchandise. I did not want to spend a large sum of money on sports apparel with a logo that no one at home would recognize so I gravitated toward the shelf containing T-shirts. Because I am a bit of a rebel, I immediately was attracted to a bright orange shirt that pictured a man with his arm raised and his index finger extended. The figure looked like someone from an old Soviet-era poster used to rally the working class. Under the figure were some words in Cyrillic Ukrainian followed by #1. I have to assume it translates to "Shakhtar football club is #1." When I wear this T-shirt my wife says, "You only wear that to draw attention to yourself." I say, "Yes, all of the T-shirts I buy and wear are intended to draw attention. They are a fashion statement." Well, the associates who walked in with UAW T-shirts were not trying to make a fashion statement. They were making a work culture statement.

To me, the UAW T-shirts symbolized a culture of low trust. Having to advertise the fact that you are a union member at work tells me there is a trust gap between management and the union. This is not unusual in union plants. To me it signaled a wonderful opportunity to make a difference by building trust during the two-day workshop. After all, I had been hired to develop all of the attendees into safety coaches who could make a safety difference in their workplace.

At the first work center at which we stopped on our end of day Safety Gemba Walk, I approached an associate who was operating a machine, introduced myself, and explained why we were going to observe him working. As he began his work tasks, we witnessed him having to physically support a long heavy metal part while it was being gripped by a machine for processing. I asked him if his back ever hurt. He looked at me with a bit of an attitude and said no. As we watched and I continued to ask questions, he seemed annoyed. The union culture was being displayed. As he continued to work, I asked him if there was one thing that could be done to make his work safer and easier. He thought for a minute and then said he had to carry the parts quite a distance to put them onto a roller bed after they were completed. He said, "It would be nice if I had a cart." I told him we would add the item to our opportunity log and we moved on.

This same attitude of noninterest in possible solutions to work practices we observed was displayed by others on our walk. Then we would find someone who was wide open to change and had numerous ideas of his or her own once we opened the door. During this walk, I closely observed the UAW T-shirted members of our team. What they observed from me was a genuine interest in making everyone's job safer and easier. I pulled them into the discussions regarding improvement ideas so they were not just observers but actual participants.

After our walk and before I departed, each of the trainees had to commit to his or her own personal Lean Safety action plan. Then when I returned in two weeks, they would have to get up in front of all of us and report out on the results of their activity. They were given the following four options from which to select, each of which had been touched upon during the workshop that day.

1. Commit to a number of Lean Safety walks during which you will engage someone else to help him or her understand how to make work safer and easier.
2. Commit to performing a specific safety standard work task.
3. Reflect on your current company safety program and recommend changes that would help to ensure that there are continuous improvement elements.
4. Create your own idea to impact the safety culture.

Most selected number one and committed to three events over the next two weeks.

When I returned after the two weeks had passed, we gathered in the meeting room and one by one they presented their stories. Their presentations highlighted positive experiences in which they had engaged their peers in Lean Safety improvement discussions. As far as the union members, they displayed the most positive of attitudes. They were friendly and open and provided very positive feedback on their interactions with their peers when they conducted their Safety Gemba Walks. They were engaged and my definition of an engaged employee is one who feels like he or she has made a difference in the business.

Before I departed, I asked each workshop participant to complete a short survey on the workshop. In response to the question, "What part of the workshop was most beneficial to you?" they said:

- Gemba Walk was great. Seeing how others work and coming up with ideas to improve their work environment.
- Safety kaizen on Monday afternoon. It showed me some new ideas on how to solve some of the problems.
- Seeing how we could improve working conditions on Safety Gemba Walk.
- Seeing all of the changes for the better that we could make.
- Safety Gemba Walk: it was the best overall chance to improve several people's lives here and at home.
- Safety Gemba Walk. Actually seeing all the things that could be improved.
- Safety Gemba Walk because it brought real-life meaning to exercises.
- Safety Gemba Walk. A great opportunity to make a difference on the shop floor.
- The Safety Gemba Walk in shipping—my area to manage.
- Safety Gemba Walk.

That is a strong recommendation for Safety Gemba Walks. My work was done. The safety culture had been impacted and the gap between the union and management had been narrowed because they had a common focus—making work safer and easier in their facility.

Case Study 16

Custom Field Service Vehicle Plant

Stellar Industries, Inc., Garner, Iowa

> "The leader's role is not to control people or stay on top of things, but rather to guide, energize, and excite."

This site was a member of a consortium and the consortium facilitator arranged my visit. I was scheduled to spend a half day on site and the schedule included a Lean Safety Gemba Walk, a meeting with leadership, and a meeting with the employee-based safety committee. This site was spread across multiple buildings, some connected and some not. This is often typical of small businesses located in rural America that either grow the existing business or bring in new product offerings. Service trucks were manufactured at this site. A pickup truck frame with cab was purchased and then they fabricated and installed a sheet metal structure with cabinets and doors onto the frame behind the cab. The finished vehicle was a field service vehicle that could be utilized by a wide variety of technical companies.

In the initial meeting, I shared my Lean Safety philosophy and answered any questions they had. Noteworthy is that some of the staff had attended the first day of my Lean Safety workshop the prior day. They philosophically understood Lean Safety and now it was time to put the theory into practice on their Gemba.

We visited the assembly area where the fabricated structure was married to the truck's frame. One of the first associates we observed was under a truck, which was positioned overhead on a hydraulic lift. He was using wire ties to attach a wiring harness to the metal structure of the truck's frame. It was dark and he could not stand completely erect so his neck and head were bent to one side (out of neutral). Because the wire ties were black, he had difficulty seeing as he pushed the leading edge through the steel structure and grabbing it so he could insert it into the receptor end of the tie. I asked him if white wire ties were available for purchase. He said he did not know. I responded, "If they were and you were using them right now, would it be easier to see the ends?" "Yes, I guess so," was his response. Today you do not have to wonder about anything, you just have to search the Internet. Someone did from their smart phone and confirmed that we could add "order white wire ties" versus "investigate the availability of white wire ties" on our opportunity log.

I was informed by the internal staff that at the next workstation, the operation we were going to observe—installing a roller drawer cabinet frame into an opening in the sheet metal structure—was known to be a difficult task. "Great," I said, "let's see what we can do to make it easier." The cabinet with drawers installed was delivered to the workstation on a pallet. The first step for the associate was to bend over and remove the drawers. A note was added on opportunity log: Have the production cell that produces the cabinet and drawers deliver them unassembled. He then lifted the frame, which was now rather unstable without the drawers, and inserted it into the opening. Next, he had to position the frame in the final position and, with a pencil inserted into the predrilled holes in the frame, mark the sheet metal structure to which it would be secured with sheet metal screws. He then removed the frame from the opening and, using a hand-held drill, attempted to drill into the base and side of the sheet metal structure. To do this he had to bend and insert his arms and head into the opening. He really struggled to complete this task. I asked him to stop for a minute and engaged everyone in a discussion on process improvement. We talked about:

- Having the vehicle on a hydraulic lift so the work area could be accessed without bending
- Adding some sheet metal straps to the frame so it was stable without the drawers
- Having predrilled holes in the sheet metal opening to eliminate the requirement to use a hand drill
- Using preformed slots and tabs in the sheet metal opening that would align and hold the cabinet in place so that it would self-align and reduce the number of screws required

Hearing our discussion, the associate jumped right in and contributed his ideas. My facilitation style is to get the conversation going by spoon-feeding the Gemba Walkers a few ideas, and then stand back and watch them explore the ideas and add their own as they dig deeper. In this case, I observed the operations manager as his engineers, supervisors, and assembly associates enthusiastically discussed how they could make the work safer and easier and reduce the cycle time to install the cabinet into the truck body. A common focus is all you ever need to build a team and this team was on a roll. After our walk ended, we returned to our meeting room where we compiled and reviewed all of the opportunities that had been identified by the Safety Gemba Walkers. Very importantly, individuals volunteered or were assigned responsibility to follow-up on some of the important tasks. After a short break, my meeting with the internal employee safety committee began.

As they filtered into the room, I observed their demeanor. Some appeared a bit nervous and others were relaxed and quickly joined in some small talk as we waited for everyone to arrive. To start, we went around the room to identify ourselves. In addition to their names, I asked everyone to talk about the work they performed and their time and role on the safety committee. I then gave them an overview of Lean Safety and what we had just discovered on our Lean Safety Gemba Walk. I then invited them to talk about their safety committee, its successes and struggles, and to ask me any questions they might have. At first, there was nothing but silence. Then they started to open up. They admitted it was a struggle to get people to stay focused on safety. When they entered a new department on a safety walk, a secret word or whistle preceded them to alert all to their presence. They said they felt like the safety police. This is a common theme expressed by safety professionals who focus on compliance safety. I offered suggestions based on my experience:

- Invite the individual who seems to be most opposed to your safety activities to join you on the compliance safety walk or invite that person to join the safety committee.
- In each area you visit, invite an associate who works there to participate in the compliance audit of the area.
- Invite the operations director to join you for your compliance walks.
- Ask individuals to express their safety concerns to you—engage as many people as you can.

■ Start a continuous improvement safety program. For instance, develop a safety improvement program that will allow anyone to suggest and implement a safety improvement in their work area. Have their participation in this program added to annual review documents in your business so they get credit for participation.

■ Offer to help co-workers find safety improvements so they see you as helpful versus the "safety police."

As I talked, I could see some of them were uncomfortable with what I was suggesting. Clearly, their boundaries and responsibilities needed to be clarified by the site leadership. Often, employee-based safety committee members do not understand the power they have. Management would like nothing better than if they stepped up, took ownership of safety in their facility, and engaged their peers to do the same. If the employees are not engaged, it will always be management's or the safety committee's safety program. The goal is to make it everyone's safety program and that will only happen if you engage a broad number of associates in the program. Having an employee-based safety committee put this company ahead of many others. Now management needed to clarify and expand their roles so they felt empowered to make a greater safety difference.

Machining and Assembly Operation, Canada

"You may never know what results come of your actions, but if you do nothing there will be no result."

—Ghandi

I was surprised when I received an inquiry to conduct a Lean Safety workshop in western Canada because the inquiry originated from the local American Society for Quality (ASQ) chapter. I was even more surprised when I found out they had never before brought in a paid presenter and hosted a workshop. I was wondering why a quality-centered organization invited me, the Lean Safety guy! Shouldn't they invite a quality guru? A few months after the original inquiry, I was making travel plans for an area of Canada I had never before visited. A year earlier, I conducted one-day workshops in both Regina and Saskatoon, Saskatchewan. The topography of Saskatchewan is a very flat prairie landscape. It is so flat you can watch your dog run away for three days! This area, on the other hand, was just east of the Canadian Rockies and provided some magnificent views of the snow-covered mountains.

The first day of the workshop was conducted in a hotel. I always prefer to hold both days of a two-day workshop at the host company site so that we can take the initial Gemba Walk at the end of Day 1. In this case, the site did not have meeting space that would accommodate us so at the end of Day 1 everyone was directed to meet at the manufacturing site the next morning.

The product produced at this site supported the oil industry. If you have not heard, western Canada is experiencing an oil boom so the plant was busy with customer orders. After the site staff gave us an overview of their business and products, we were given the required safety apparel and headed out for our initial Safety Gemba Walk. The site representatives explained their production processes as we walked the flow of their products. Whenever I recognized an opportunity, I stopped to engage production associates and the Gemba Walkers in discussions. Simple things that initially seem insignificant can be the trigger point that opens a person's eyes to the true meaning of Lean Safety. An example that occurred on our walk began when I observed an associate inserting a large part into a test machine. After initiating the controls to start the test, he had to observe a gauge to determine when the test process had concluded. What I observed, and what others may have missed, was that the controls and gauge were around the side of the machine, which caused the operator to lean around the machine to do his job. When I asked the workshop attendee next to me what he had seen as we watched this operation, he seemed puzzled. When I suggested the controls and gauge should be relocated to the front of the machine to prevent the out of neutral movement, he said, "I don't understand how you see those things." Practice, practice, practice was my response—let's keep walking. After we concluded our walk, the attendees were split into two safety kaizen teams and assigned different work areas where they were to engage the associates in safety improvement discussions.

One of the safety kaizen teams was sent to a finished goods packaging area. I joined them for their initial observation of a current state packaging process. Large steel shafts were delivered to the packaging area on pallets. They were then wrapped with protective material before being inserted into wooden crates for shipment. When the team engaged the associates in a discussion, we discovered they were temporary employees. We explained our objective, to make work safer and easier, and the role they could play by both helping us understand the work process and giving us their improvement ideas. Even temporary employees can be engaged when someone says they want to make their job safer and easier. They began by both bending over—one to lift one end of the heavy shaft while the other inserted a nylon cable sling under the shaft to locate it on the center of the shaft. One end of the sling was inserted through the loop on the other end to cinch it around the shaft. The free loop was then attached to the hoist hook. Because they only used one cable sling, the load was not well balanced and one of them had to physically support one end as the shaft was moved with the hoist toward the shipping crate. Then they would lay a piece of protective paper material inside the crate and lower the shaft. Because it was a tight fit in the crate, one of them would bend over and lift one end of the shaft as the lower end was inserted first and then the end being held was guided into the crate. One of them then had to bend over to lift one end back up so they could remove the cable sling. They then bent over to wrap the center section of the shaft with the protective material. The team asked the following questions:

- Why are the shafts lying flat on a pallet?
- Shouldn't there be "V" blocks under the shafts to prevent them from rolling?
- Why is one sling used to lift the shaft?
- Why not use two slings on the smaller ends of the shaft to balance the load and eliminate the lifting of the shaft to install the sling?
- Why is the protective paper inserted into the crate?
- If the shaft were balanced using two slings, could it be wrapped with the protective coating while it is still attached to the hoist?
- If it is balanced on the hoist, could the shaft simply be lowered into the crate without bending and handling it? Could the sling now be slipped off the ends of the shaft without lifting it?

What happened next was amazing. The kaizen team left the shipping area and headed to our meeting room for lunch. About 45 minutes later when they returned, the two temporary associates along with the shipping supervisor demonstrated the new process utilizing all of the ideas that had been suggested and discussed before lunch. The team couldn't believe it. I was now with the second safety kaizen team when one of them approached me. He asked if I could join them for a minute. As we all watched, the two temporary associates demonstrated their newly defined safer and easier work process. They were beaming with pride and so was the safety kaizen team. A side benefit was the fact that two people were no longer required for this work task. That is Lean Safety—two beneficial outcomes, safety and cycle time.

Case Study 18

Sheet Metal Products Plant, Canada

"You can observe a lot just by watching."

—Yogi Berra

The Canadian province of Manitoba lies on the eastern edge of the Canadian prairie and is as flat as the cornfields here in Illinois where I live. An interesting fact from Wikipedia is that "the generally flat terrain and the poor drainage of the Red River Valley's clay-based soil results in a seasonal explosion of insects, especially mosquitoes."[*] Insects were not a worry for me because I visited in mid-November. The winds that blow across western Canada are funneled into this province and, based on my experience, are then channeled right down the road on which I walked each night to get to a restaurant. I am from Chicago area and Chicago is often referred to as the "Windy City." This moniker refers to a period in Chicago's history that was dominated by blow-hard politicians and braggarts rather than weather-related wind. Despite that I think Chicago should relinquish the title to this area of Canada! However, despite the cold winds both the people who invited me to conduct the workshop, Canadian Manufacturers and Exporters (CME), and those at the host company greeted me warmly and made me feel very welcome.

This was a two-day workshop attended by members of a local consortium that is facilitated by the CME office professionals who reside in the area. The products produced at this site were made from sheet metal. Many

---

[*] http://en.wikipedia.org/wiki/Winnipeg

of these parts, after they were cut and formed, went through a paint line. One of the safety kaizen teams was assigned this area to observe on Day 2 of the workshop. I joined them at the start of the day to observe and ask a few questions. Most of the activity on any paint line is where the parts are loaded for painting and where they are removed after painting. This team began their search for improvements at the loading station. The first opportunity they discussed with the associates working in the area related to the many different carts loaded with inventory that were crammed into a tight area while waiting to be loaded onto the paint line. They discussed having the inventory set into flow lanes so the parts waiting to be painted would automatically be sequenced. They also observed how the parts were put onto the carts and the effort it took for the associate to remove them. They noted on their opportunity log that orienting the parts in a particular direction when loading in the production area would reduce the out of neutral body movements of the associates on the paint line. They also explored a new layout for the area that would minimize the walking back and forth they observe.

Finally, to get them to focus more on the associates performing the work tasks, I asked them what they saw when parts were loaded onto the paint line. For each part, the associate had to hang the parts on a wire hook that was strung from the conveyor system that moved the parts through the paint booth. They quickly noticed the associates sometimes struggled to get the part onto the hook. They had their shoulders out of neutral and were walking to keep pace with the moving conveyor while trying to mate the hook end with a hole stamped in each part specifically to accept the hook. The associates pointed out to the safety kaizen team that the hooks are coated with paint each trip through the booth and eventually they have to remove the hooks and soak them in solvent to remove the paint. If they did not do this, the diameter of the wire hook grows due to the paint buildup and they cannot mate the two parts. I then asked, "Why is the hole that diameter?" Everyone looked at me a bit puzzled. If the hole is only there so you can put it on a hook for painting, then why not increase the size of the hole so it is easier to hook the parts? Off to engineering went two members of the kaizen team who worked at this site. They returned in a short while with drawings for some of the parts and explained that the clearance between the wire diameter and the stamped hole was only .032 thousands of an inch. They said the diameter of the hole could be increased to provide at least twice that amount of clearance. When they shared the news with the associates who hooked parts all day long, there were some smiles. They then grabbed a few parts and drilled the holes to enlarge them to the proposed new diameter so they could test them on the hooks. The difference was amazing. Someone said, "Heck, I can hook these parts blindfolded!"

Year after year, this company made existing products and designed many new ones. On every part that required painting, the engineer selected the same diameter hole for paint hook insertion. Meanwhile on the paint line, they struggled for years mating the parts to the hooks and had to spend time soaking and scrapping hooks to remove the paint buildup. So much waste because no one questioned, or believed they had the authority to question, the diameter of the hole. This is a great example of how small changes can make a huge difference for those who work with their hands.

During the late afternoon team report out to management meeting, the internal staff couldn't wait to share the proposed change to the diameter of the stamped hole. This small change meant a lot to them because it had validated the benefit of this workshop. Some weeks later, I received the event survey results from my CME contact. One of the responses to the question, "What was the most beneficial outcome of this workshop?" was simply "The hole size change." Someone who had not participated in the workshop would be scratching his or her head while trying to understand that response. To me, it made perfect sense.

Case Study 19

Beverage Can Operation, United Kingdom

"Honest criticism is hard to take, particularly from a relative, a
friend, an acquaintance, or a stranger."

—Franklin P. Jones

An acquaintance from England organized this workshop. He manages conferences and other events sponsored by his employer, the publisher of both
the *Lean Manufacturing Journal* and *The Manufacturer*. *The Manufacturer*
is the preeminent manufacturing journal in the UK. It is comparable to
Industry Week in the United States. My relationship with the organizer
goes back quite a few years. When I was still working in the industry, the
company for which I worked had purchased a small company in Glossop,
Derbyshire, England. I had the opportunity to facilitate a couple of continuous improvement events on site. Before departing for the first visit, I emailed
some UK Lean professionals who had attended the AME international conference the prior year. I asked if they could provide a contact name at a
company in the Manchester area that had a viable Lean effort in place. My
goal was to plan a benchmarking visit for some of the Glossop site staff.
The person who provided that contact name managed the Manufacturing
Institute in Manchester. She asked if we could meet and talk when I made
that first trip to England. I did meet with her at the airport hotel prior to
my flight home. We had a great conversation and early the next year I
was asked if I could present at a conference they were planning. I agreed
and it was at that time I met my current contact. He was the marketing

professional who organized the conference. We have maintained close contact for at least eight years and he is a great supporter of and believer in the Lean Safety concepts. His network is broad and he has provided me multiple opportunities to share my Lean Safety story. I cannot stress enough the importance of having a strong network if you are an independent consultant as I am. There are literally thousands of Lean consultants and without good contacts who understand the value you deliver, your reach is very limited. In addition, without his help my wife would not have felt as if she stayed at Downton Abbey!

Although Day 2 of the workshop was held in a large manufacturing site that produced aluminum cans for the beverage industry, Day 1 was held in a former manor house. The site had been converted into a hotel with a golf course now occupying the former manor grounds. The very popular BBC television series *Downton Abbey* had just finished its first season on PBS in the United States. My wife was a big fan, so I ended up taking more than one picture of her at the entrance of her temporary manor house. When the organizer and I requested that she staff the registration desk on the first morning of the workshop, she appeared a bit insulted—as if this work was below the lady of the house!

Since we were in a hotel for Day 1, our initial tour of the plant did not occur until the second morning. As you might imagine, the production of aluminum cans is a highly automated and very high-speed process. The site representatives who were participating in the entire workshop guided us around. Since the title of my book and the workshops I present is *Lean Safety,* the workshops always attract a cross-functional mixed audience. There will be seasoned compliance safety professionals mixed with operational and Lean professionals. Lean professionals are in tune to observing for improvement while safety professionals have been trained to observe for noncompliance. I have to pay closer attention to the safety professionals when I am pointed out examples of changes that would make work safer and easier. As I ask questions and point out those examples to engage the Gemba walkers, I often notice the safety professionals are in their own little world. They are often lost in the opportunity to find compliance problems at the site they are visiting. They cannot help themselves—this is their expertise and they want to add value using their individual skills. I have seen this often enough, and based on their feedback at the report out sessions that conclude the second day's workshop activities, to know one of the reasons this happens is they rarely get the opportunity to visit other facilities. They always acknowledge the host site and thank them for allowing them to come in, meet other safety professionals, and contribute to the event. They just love the networking aspect of the Lean Safety workshop. The reason this is not a routine or even an occasional activity is no company wants to have outsiders come in and point out its safety "dirty laundry." I and others who make the first inquiry to understand if a site is willing to host a Lean Safety workshop often detect this same reluctance. It is only after explaining that the workshop attendees will not be focused on compliance safety that people really start to open up and listen. In reality, because the safety professionals cannot help themselves, our Gemba Walks always harvest a mix of compliance and Lean Safety improvement opportunities. So it was on this walk.

We began where the beverage can blanks were stamped from strip stock before being formed into a can shape. This was one of those highly automated plants that required you to search for someone physically working. It was easy to be caught up in watching the cans as they moved through the automated process. It was my first exposure to this type of industry and I found it fascinating because of the volume produced and the speed of the production. Our first opportunity to observe someone working was on a raised platform where the machines formed the blanks into a can shape. There were multiple machines making parts that were discharged onto a wide conveyor. The associate had to adjust and set up the machines as needed to keep them operational. Based on my work history, I was exposed to nail makers and cold headers. Those machines take a piece of wire or rod (think of large-diameter wire) and move it across multiple forming stations to produce a finished part like a nail or a bolt. The equipment we were observing was similar in that they were about waist height and to work on them you had to lean in and over the machine. If you have ever worked in the engine compartment of an automobile, you know what I mean. You end up laying on one of the front fenders to take the strain off your back. Just a few weeks ago, I presented the Lean Safety story to a local Lean Construction Institute (LCI) meeting held in Chicago. Of course, I talked about out of neutral body positions and the effect that has on individuals. Shortly after I finished, someone approached me from the audience. He was rather excited and had something he wanted to show me. It was something he invented that would allow an auto mechanic to have his back out of neutral without risk. It was a tubular steel device that was similar to a kneeler at church except the foam-padded surface where your knees rested was higher. Once in that position, the mechanic could lean forward and rest his upper torso on the upper portion of the device, a padded platform that extended over the fender. The unit was on spring-loaded wheels, so it could easily be positioned for use. He noted his attempts to market his invention to auto dealerships, where multiple mechanics are employed, had not been going well because they were probably more interested in watching their costs than the mechanics' ease of work. Individual mechanics who have tested the device really like it. I suggested he time mechanics performing the same task, say changing spark plugs, with and without his device. The cycle time difference, along with any back strain or injury history, should then be a compelling selling point he could use when approaching the dealerships.* I remember that as I observed the associate lean over his can-forming equipment, I pondered how to take the strain off his back. The invention I just described could be easily adapted for this industrial use. It is nice to know others, without even knowing it, are pursuing Lean Safety improvements.

One of the primary weapons local producers have in their competition with overseas producers is their proximity to the customer. This site took advantage of that competitive advantage by co-locating their facility extremely close to one of their largest customers. They were so close that if you wanted a soft drink you could walk next door and get a freshly bottled can. Only a wall separated the can-making site from the bottling plant. A conveyor fed the newly made cans directly to the production line of the bottling plant. Empty aluminum cans are containers of air and both facilities understood the waste inherent in loading them onto trucks and paying to transport them anywhere. By setting up a production process that improved the flow of products from supplier to customer, they also eliminated all of the safety risks associated with loading, transporting, and unloading pallets of goods. What a great example of Lean Safety—cycle time and safety gains resulting from one improvement effort.

---

* www.kreep-up.com

Case Study 20

Process Production Operation, Australia

"In the beginner's mind there are many possibilities, but in the expert's there are few."

—Shunryu Suzuki

This was an in-house workshop with cross-functional representatives from multiple corporate sites. Their EHS manager had attended a two-day Lean Safety workshop about six months earlier. At that event and in follow-up email exchanges, she admitted to struggling with understanding how the principles of Lean Safety would apply to her process-focused business. Therefore, it was a bit of a surprise when I was informed they had requested to be the site of a Lean Safety workshop.

During our initial Lean Safety Gemba Walk, at the start of Day 2 of the workshop, it took us close to 45 minutes to find someone working. This is the Gemba Walk during which I, the Lean Safety expert, am supposed to teach the participants to see safety differently as we observe people working. Because this facility, like many process plants, runs on automatic, people often are not engaged in the production process itself. Instead, the majority of them perform technical and maintenance related tasks that help to ensure the process plant is running as it is intended or scrambling to get it going again when problems occur. I was a bit anxious when 15 minutes later I turned three teams loose to identify a targeted 100 Lean Safety improvement opportunities. However, if there is one thing I have learned it is to trust the process. No matter how concerned I am, the teams always do a great job of finding plenty of improvement opportunities.

One of the teams focused on an area of the facility that was not included in our initial Safety Gemba Walk—the quality assurance laboratory. Because the product manufactured at this site had many variations, there were stringent requirements to test both the raw materials and the finished products. The lab, I was told, had been targeted for an upgrade. Their plans were to transform the lab, which currently had a worn-out vintage 1970s look, into something modern that would be a showplace for customer visits. I had seen a similar transformation at a Chicago-area company a few years prior.

The lab I am referring to performed testing on both the components for and their finished products that were designed for the oil and gas industry. It also had that 1970s look that helps to keep young, talented people from pursuing manufacturing careers when they are given plant tours during their job interviews. Even worse is when customers are brought into these historically uninteresting settings that do anything but create a sense of confidence in the supplier. This company, or maybe their sales and marketing team, understood the importance of this customer touch point and dedicated some capital budget to an overhaul of the lab. Assigned the task was a young manufacturing engineer who had recently been influenced by another manufacturing plant in the Chicago area. Together, as volunteers, we had visited the facility to ask if they would consider being a tour site for a large Lean conference the following year. The owner of this firm is not only a patron of the arts, he believes manufacturing is an art. Within the plant are art, an art gallery, and a unique culture I have seen nowhere else. Art is embedded as deeply into this business as accounting is in every business. The facility is the opposite of what people think manufacturing plants look like. They have taken art and made it the focal point of the business, the thing you will never forget about their business after your visit. Visit their website (http://www.winzelergear.com/) and you will understand what I mean.

After viewing the artwork in this plant, the youthful engineer decided their lab overhaul should consist of more than an upgrade of storage cabinets, workbenches, and chairs. She was determined to make their lab a showplace that would both elicit pride from the people who worked there and impress their customers. As is often the case, what one person sees as an imperative may be viewed differently by the people who actually have to live with the change. Those who worked in this lab displayed some resistance, or at least apprehension, to a proposed lab design that included brightly colored walls on which would be displayed large format photographs that were skillfully enhanced digital renditions of the company's products. Just because there is pushback does not mean new ideas do not have merit. It only means you have to work hard to help people see the value you perceive in the proposed change. In this case, the resistance was overcome and the resulting lab was indeed a source of pride for everyone who worked there.

In the process plant lab, the safety kaizen team and I observed an employee who was making up test samples by mixing, molding, and then curing them in an oven. I had been forewarned that this employee could be difficult and might be resistant to any ideas for improvement. I love it when I hear an employee I am about to engage in a discussion about making his work safer and easier may be difficult. I love it because I know I have been given a chance to change how someone thinks, acts, and interacts (the business culture) and in turn how people view that individual. He seemed a bit agitated as we watched him work in a process that had him walking almost in circles and for long distances to complete his work task. He was working at a rapid rate to either try to impress us or because he was irritated that we were there observing him. In Lean terms, there was poor workflow because of the equipment layout. Historically, work area layouts were defined by management, and then employees, who had no input into defining the layout, are asked or even worse told to work there. That approach has to change so that management is doing it with them, not to them. When I asked him if anything could be changed to make the work easier, he quickly pointed out some manual handling issues. As soon as a person gives you an entry point, you can begin to start a dialogue about improvement, which I did. In his somewhat animated responses he often used the terms "them" and "they" when referring to management. I could tell there was a lack of trust in the relationship and this desire by management to change and upgrade the lab was a perfect opportunity to begin trust building. That is one of the benefits of being a consultant, or outsider. I am not involved in the politics and broken relationships that exist in a business. I am neutral and that allows me to ask simple questions in an adult-to-adult manner. I suggested that to accomplish the task of defining a new lab layout, a kaizen blitz event could be scheduled to define a new workflow that would reduce the risk of soft tissue injuries. I would invite this perceived difficult employee to be on the team. Then using block and relationship diagrams, and ultimately relocating the equipment, a workflow would be defined that would allow the operator to work at a steady pace and keep his body parts in neutral while he performed the work tasks. Outcomes of this kaizen effort would be improved relationships, a new layout that would reduce injury risks and work process cycle times, and the freeing up of some valuable lab space. The approach could then be taken to define the layout for the other sections of the lab. Obvious to all was the fact that not one cent should be spent on the lab upgrade until the kaizen events yielded those results.

Management teams who have embarked on the continuous improvement journey have to earn their employees' trust one person at a time, and here was an excellent opportunity to change one person's view of management. A perceived difficult employee who changes his or her behaviors is the best salesperson a company on the Lean journey can have. I think this individual was a frustrated employee, which caused others to see him as difficult to deal with. I was informed that when he had been asked what changes he would like to see in the lab his response was "Everything is okay as it is." Today's workers want to be engaged and have a say in their work life and I believe he did also. The relationship history prevented him from honestly participating in the change discussions. I believe he felt he had not been heard in the past and was therefore reluctant to contribute. I only wish I could have stayed to engage him in the change process by facilitating the kaizen events. I have no doubt I could have brought him over to the "Lean" side by collaborating with him to make his job safer and easier. I have yet to come across an individual who has balked at that offer.

Conclusion

"Maybe the ultimate test of a good leader is whether they have the ability to change the way people think. It would seem to me that the final obstacle between winning and losing is to get your team to think differently about a problem, or a product, or a strategy; to guide them to seeing things in new ways; and then getting out of their way, knowing that once there is an awakening within, the leader has been successful."

—Joe Zeno

My reason for writing this book is quite simple. I want the readers to take the case studies to heart, learn from them, and go make a difference wherever they work or have influence. In order to individually make a difference, they must understand, believe in, and practice people-centric leadership. People-centric Lean leaders are driven by the inherent value of every individual and have a clear understanding that everyone wants to and can become more than he or she is today. The impact people-centric Lean leaders can have on individuals in their charge can be amazing. Here is an example that I witnessed while facilitating a session in the Idea Exchange Café at the AME International Conference. The topic open for discussion in the first session of the day was "Methods Used to Engage Employees in Continuous Improvement." Two of the session attendees were from Africa. They were part of an African contingent numbering around 40 people who represented a variety of companies. One of them, Dickson, seemed to me to be a bit nervous. He was restrained and did not readily engage in the discussions that were taking place. When he finally opened up, the reason became very clear. He felt uncomfortable because he, a maintenance mechanic in a grain mill, had been selected to go to "America"—a very distant land in which he was not comfortable. He noted that his co-workers back in Africa were in awe of this "once in a lifetime experience" his employer had provided him. Dickson proceeded to tell us about the Lean implementation underway in his plant and the role he played. Because he had displayed an interest and a passion for Lean, he had been selected to be a Lean coordinator and was now sitting in a session in "America" sharing his experiences. He could not have imagined this experience in his wildest dreams. If anyone could have observed the expression of pride on his face as he talked about his plant, and the change processes he supported, they would have had the opportunity to truly understand what Lean is all about. It is not about cost savings; it is about growing the people in a business, any business, so that everyone can make a difference. Everyone making a difference drives business growth and differentiates your business. People-centric Lean leaders have the opportunity to touch people, like Dickson, and make a difference in their lives so that they can in turn make a difference in the business. I hope reading this book of case study stories has inspired you to go inspire others.

Reading about the philosophy of technical subjects like Lean or safety can be uninteresting and frankly boring. Using a case study framework to continue my writing on the topic of Lean Safety was intended to add interest and supports the way in which adults prefer to learn. Malcolm Knowles, an American practitioner and theorist of adult learning, identified six principles of adult learning.* They are:

- Adults are internally motivated and self-directed.
- Adults bring life experiences and knowledge to learning experiences.
- Adults are goal oriented.
- Adults are relevancy oriented.
- Adults are practical.
- Adult learners like to be respected.

Let's cover each of these bullet points to support my case.

1. **Adults are internally motivated and self-directed.** My hope is that after reading these case studies, the readers will be motivated to make a safety difference because it is the right thing to do. Compliance safety is an external motivator. I hope these Lean Safety case studies internally motivate the readers.

---

* http://www.qotfc.edu.au/resource/?page=65375

2. **Adults bring life experiences and knowledge to learning experiences.** The case studies contained in this book are some of my life experiences, and my knowledge of Lean Safety was gleaned from those and other experiences. After reading this book, I expect the readers to experientially learn and add new life experiences to their resume.

3. **Adults are goal oriented.** I hope all readers set some goals related to leading Safety Gemba Walks or Lean Safety kaizen events. Just reading the book without changes in behavior, in Lean terms, might be considered a "waste."

4. **Adults are relevancy oriented.** There is nothing more relevant than the safety of people who work in the facilities managed or supported by the readers of this book. After reading this book, I expect the readers to have a burning platform and the passion to lead real safety change that will make a difference in the everyday work life of their reports and co-workers.

5. **Adults are practical.** If that is the case, then the hands-on approach of taking Safety Gemba Walks to engage people in discussions about making work safer and easier is a perfect example of the practical application of a common sense approach to employee engagement.

6. **Adult learners like to be respected.** My goal was to respect the readers of this book by sharing real-life stories that clearly defined a path toward real, meaningful safety change in organizations. If you, the reader, in turn move away from the parent-child relationships, which result from activities of compliance safety programs, and toward the adult-to-adult interactions that occur during Safety Gemba Walks, you will earn the respect and trust of all whom you touch. Go make a safety difference.

Stay safe!

Index

About the Author

Robert B. Hafey, an operations and Lean professional, spent over 40 years working in manufacturing at U.S. Steel Corporation and Flexco. His first book, *Lean Safety— Transforming Your Safety Culture with Lean Management*, was the first to link the topics of Lean and safety. This positioned him to build a successful Lean consulting business— RBH Consulting LLC.

He firmly believes in the email signature tagline he created, "You can continuously cope, or you can continuously improve—the choice is yours!" He considers continuous improvement a creative endeavor and shares his passion for the topic wherever and whenever possible.

He resides in Homer Glen, Illinois, with his wife Sandy. They have three grown daughters, Liz, Kate, and Colleen.